Latigo Leather

Also by Herb Genfan and Lyn Taetzsch

How to Start Your Own Craft Business
Leather Decoration

Latigo Leather

BY HERB GENFAN AND LYN TAETZSCH

WATSON-GUPTILL PUBLICATIONS / NEW YORK

Copyright © 1976 by Herb Genfan and Lyn Taetzsch
First published 1976 in New York by Watson-Guptill Publications,
a division of Billboard Publications, Inc.,
One Astor Plaza, New York, N.Y. 10036

Manufactured in U.S.A.

Library of Congress Cataloging in Publication Data
Genfan, Herb.
  Latigo leather.
  Includes index.
  1. Leather work. I. Taetzsch, Lyn. II. Title.
TT290.G46   745.53'1   75-35620
ISBN 0–8230–2650–7
ISBN 0–8230–2651–5 pbk.

First printing, 1976

For Elsie and Charles Stoessel

Acknowledgments

We thank our editors at Watson-Guptill for their help in this and all our books. We also thank all of the leathercrafters who contributed photographs of their work to this book, and especially the employees of Holy Cow Leather, Newfield, New York, whose work went into many of the projects shown. We thank Mary and Braum Azerbegi of Braum's Leather in Berkeley, California, for being our first tutors in leathercraft and starting us on our way.

Photographs by Herb Genfan
Illustrations by Ilene Zetterberg

Contents

Introduction

What is latigo? One of the leading leather suppliers, MacLeather Company, defines latigo as "a type of chrome-tanned leather that can be stamped but not tooled or modeled." Another well-known supplier, California Crafts Supply, says latigo is "an oil-tanned brand-free, general-purpose cowhide used for handbags, belts, sandle straps, etc." Still another supplier defines latigo as "alum-tanned cowhide, finished with waxes and oils to produce a supple, extremely durable and weather-resistant leather that mellows with age."

So, we know that latigo is cowhide, and is not vegetable-tanned. It is a natural-looking, mellow, durable leather that can be used for all the projects in this book and many more. Outside of these qualities, latigo varies from supplier to supplier. We've seen latigo in all colors, including white (although the standard color for latigo is yellow); in thicknesses from 3/4 oz. up to 9/10 oz; with variations in surface from extremely porous and grainy to smooth, waxy, and dye-resistant; in variations of stiffness from extremely soft and pliable to firm and somewhat hard.

Latigo comes in sides, which means the hide is split in half along the spine. A side—which usually averages about 20 square feet—includes the head, shoulder, and leg areas of the hide. Leather is sold per square foot, and although most of the suppliers we've listed will sell a single side at a time, you can usually get the leather cheaper by buying in larger quantities.

This book provides the reader with patterns, simple directions, and step-by-step photographs illustrating how to make eleven projects. No previous knowledge of leathercraft is required to make the projects in this book, but even if you do have some leather experience, we're sure you'll find new ideas and methods to try.

The first chapter explains how to set up a work area; the second, the tools and supplies you'll need and how to use them. We suggest you practice the techniques shown in Chapter 2 before trying the project later on in the book.

The last chapter in the book shows how to design your own handbag patterns. Make at least one of the handbags in Chapters 12 and 13 before tackling a pattern yourself. Then once you've mastered the technique of designing handbag patterns, you can easily transfer this knowledge to making such other projects as saddle bags, quivers, etc.

Other than dyeing, we haven't covered any decorative techniques in this book. But since latigo is so receptive to many decorative materials and techniques, we suggest you read our previous book, *Leather Decoration* (Watson-Guptill Publications) for ideas in this area.

Latigo Leather

1 Getting Started

There are various kinds of latigo leather available, and in this chapter we'll discuss the differences. We'll also describe the work area and the supplies for it as well as the tools you'll need when working with latigo. The Suppliers List at the end of the book tells you where the tools can be purchased.

Varieties of Latigo

As mentioned in the Introduction, latigo may vary a good deal from one tannery to another. Most latigo comes in a shade of yellow, although you can get in in other colors and white. Yellow latigo is the easiest to work with—it is the most pliable, takes dye the best, and is most readily available. We suggest you start out with yellow latigo and experiment later with other colors.

Latigo comes in different thicknesses, too. The thickness is measured in ounces, with one ounce equaling 1/64". We suggest 8/9 or 9/10 oz. latigo for belts and 6/7 or 7/8 oz. for handbags and belt pouches. Scraps of any of these weights can be used for barrettes, keyrings, earrings, wristbands, and watchbands. Refer to the chapters on each project for specific details on the best latigo to use.

The surface of different kinds of latigo varies also. Each type dyes differently and has different degrees of flexibility. Some kinds are glazed and resist dyeing while others soak the dye up like a sponge. When you first order latigo from a supplier, try out only a small amount. By experimenting you'll come to prefer a particular type of latigo for your work.

Work Area

You'll need a sturdy table or workbench and a stool if you like to sit while working. To protect the top of the table when you're cutting, use a piece of plywood or heavy scrap leather under your work. A piece of cardboard will protect the table while you're dyeing. Make sure the lighting is adequate so you can see your work without straining.

Cutting belts from a side of leather will require a large working area, preferably the floor. To protect the floor or other surface when you make the first long cut, place a piece of 1/4" plywood or several layers of cardboard under the leather. We keep a 4' x 8' piece of plywood on the floor at all times for cutting belts and handbags.

For riveting and setting snaps you'll need a hard, smooth, solid piece of metal. We suggest placing a steel plate at least 1" thick on your worktable. We use a 2" thick, 10" x 12" steel plate, although some leathercrafters use an anvil. You can find steel plates at scrap steel yards, and anvils at shoe repair supply stores.

Tools and Materials

Now we will discuss all the tools and materials needed to complete the projects in this book. As you continue to work in leather, you will undoubtedly need a wider selection of the numerous tools available. However, although the tools and materials included here are only the basics, they can be used for many other projects.

Latigo. These pieces of leather are all latigo, although they vary in thickness, color, flexibility, and surface finish.

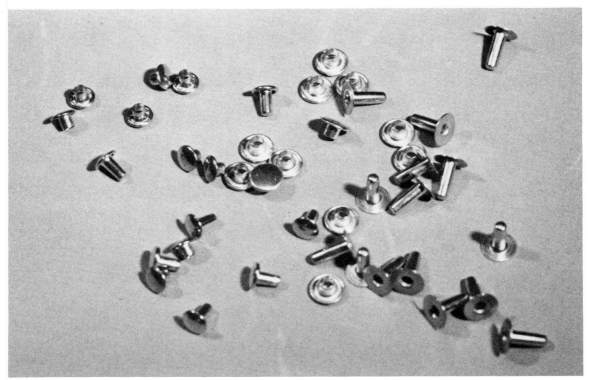

Rivets. Rivets come in different sizes and colors, and are convenient, durable fasteners that hold pieces of leather together.

Rivet Setter. Rivet setters are used for setting rivets.

Snaps. Snaps also come in different sizes and colors. They provide a simple fastening that can be opened and closed with ease.

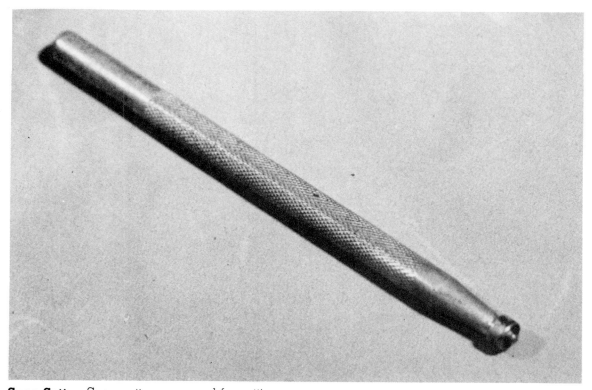

Snap Setter. Snap setters are used for setting snaps.

Buckles. Buckles come in many
sizes and shapes, and may be made
of various metals, stained glass,
leather, or other materials.

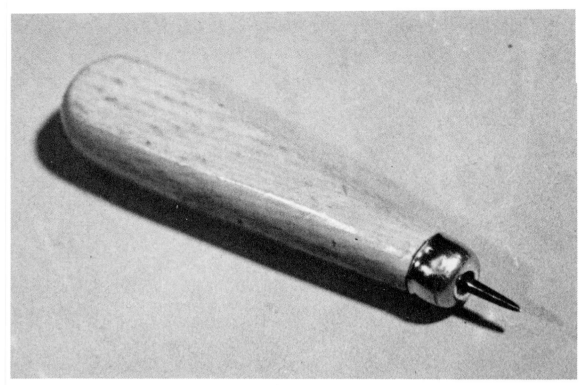

Stitching Awl. This stitching awl is used for tracing around cardboard patterns on leather, making holes, and drawing decorative lines.

Stript Ease. The Stript Ease is used to cut belts, handbag straps, and other strips of leather.

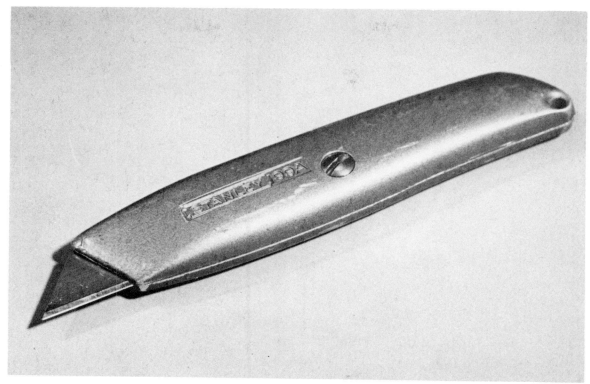

Utility or Mat Knife. This utility knife or mat knife is used for general cutting.

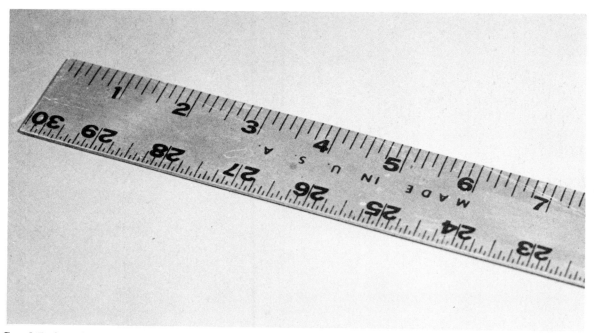

Steel Ruler. A steel ruler is used for measuring, and as a straight cutting edge.

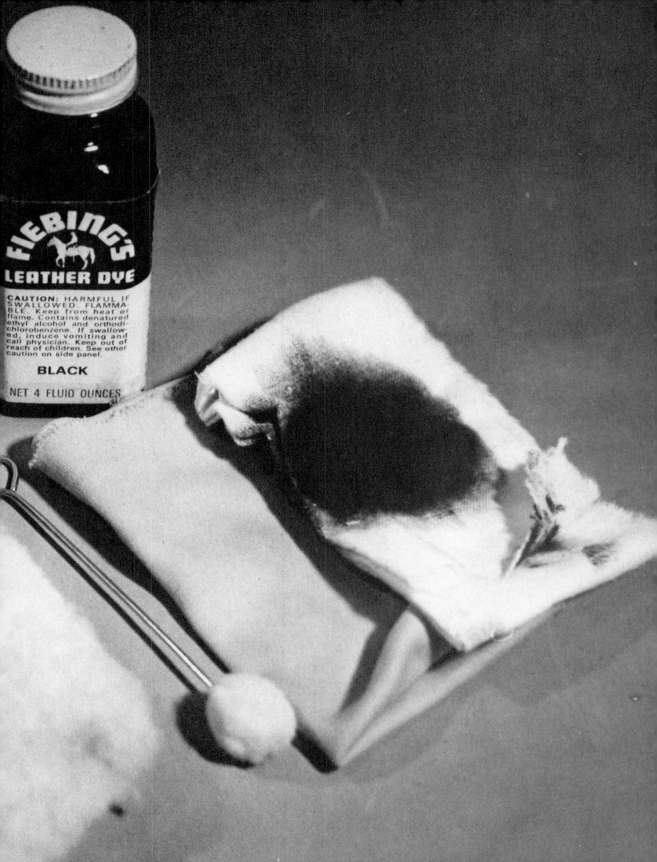

Dye. It isn't necessary to use dye on any of the projects in this book, but since most latigo leather can be dyed very effectively, we've included instructions on it. Shown here are Fiebing's dye and saddle soap, a dauber, pieces of flannel, and sheepskin scraps.

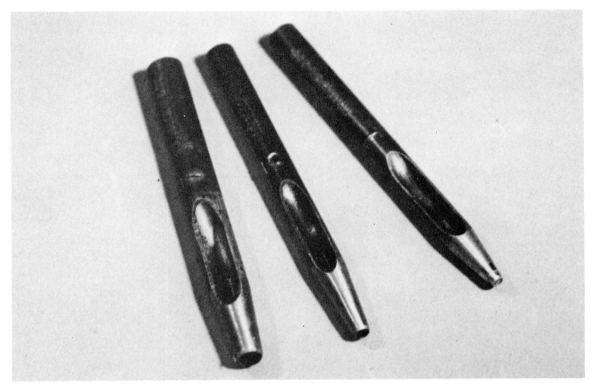

Round Hole Punches. Round hole punches come in various diameters and are used to make round holes.

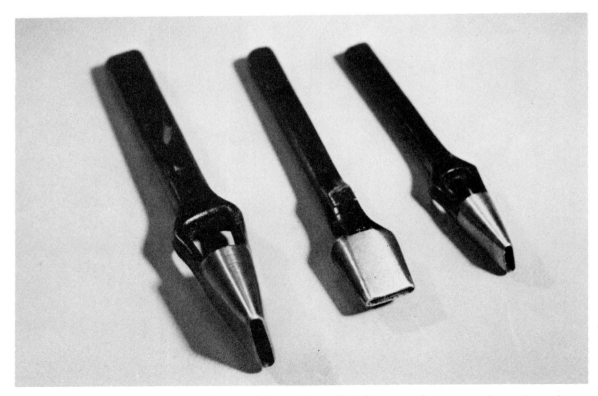

Oblong Hole Punches. Oblong hole punches, also called bag punches, come in various sizes and are used to make oblong holes in watchbands, belts, etc.

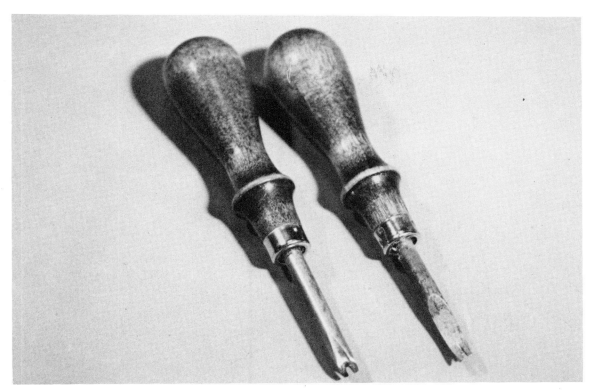

Edge Bevelers. Edge bevelers are used for beveling, or shaving at an angle, the edges of cut pieces of leather. A #2 or #3 edge beveler is usually the best size to use for latigo.

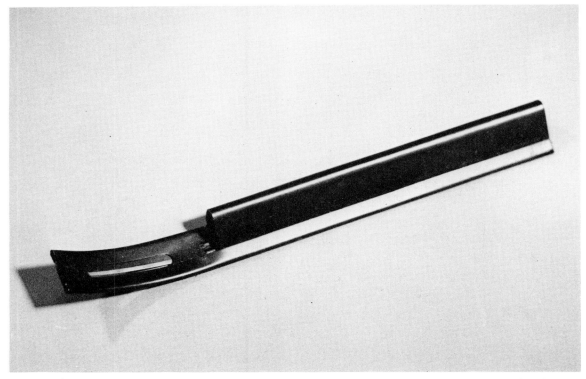

Skiving Knife. A skiving knife is used for shaving or thinning down a piece of heavy leather.

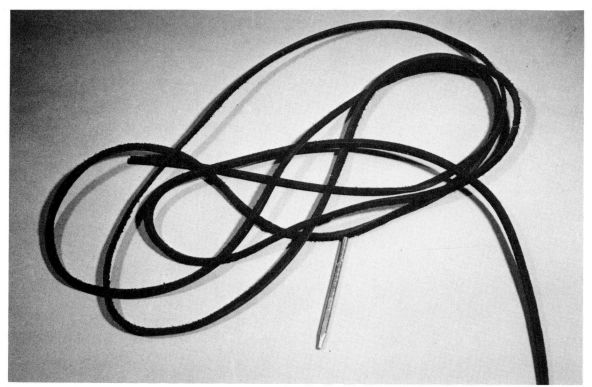

Latigo Life-Eye Lacing Needle. This is used with ⅛" latigo lace.

Rawhide Mallets. Rawhide mallets come in several sizes and are used on the round and oblong hole punches, the rivet setter, and the snap setter.

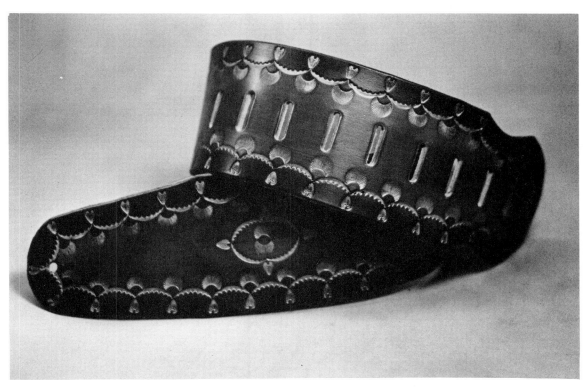

This guitar strap was decorated with stamping tools and dye. By John Cowan of Ithaca, N.Y.

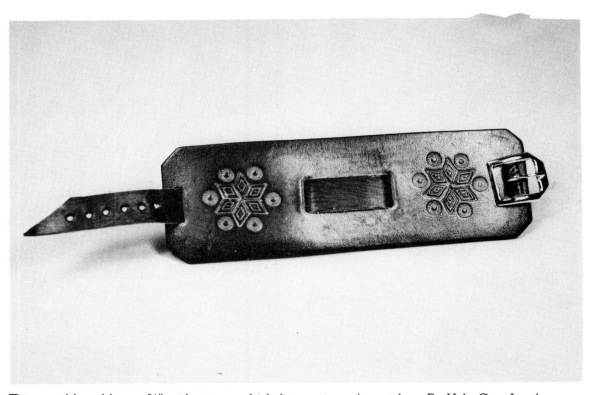

This watchband has a ⅝″ wide strap, which fits most men's watches. By Holy Cow Leather.

2 Basic Procedures

In this chapter we'll demonstrate how to use the tools and materials necessary for the latigo projects in the following chapters. There will be a description of how to use each tool along with a photograph illustrating the proper technique. We suggest you practice these procedures on pieces of scrap leather first before beginning a project.

For specific information on attaching buckles to watchbands and belts, see Chapters 7 and 8. For instructions on lacing latigo leather, see Chapters 9, 12, and 13.

Tracing, Marking, and Cutting

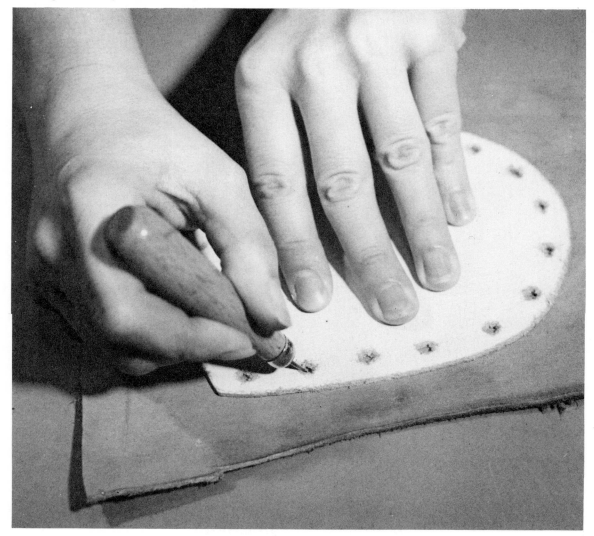

1. Use an awl to trace the cardboard patterns on leather and to mark each hole.

2. To cut out leather pieces, use your utility knife. Make sure your blade is sharp, especially when cutting curves.

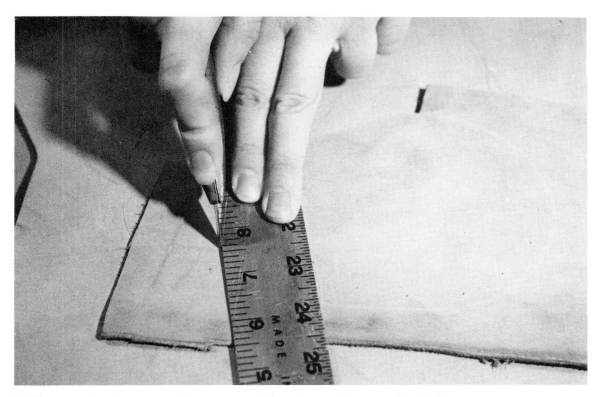

3. Use a steel ruler as a guide to cut straight edges with your utility knife.

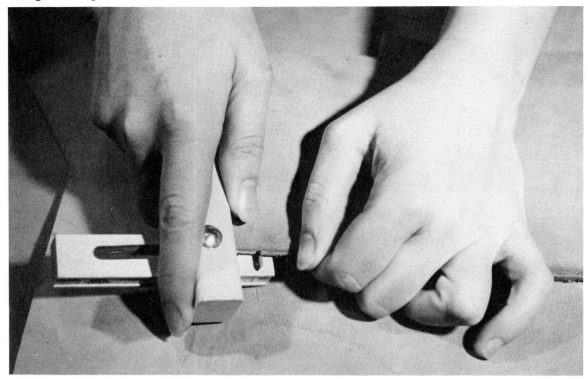

1. After cutting a straight edge on the right-hand side of the leather, cut strips with a Stript Ease. Hold the leather edge firmly against the wood of the Stript Ease.

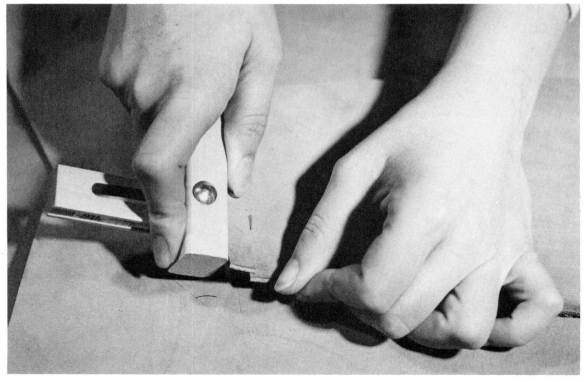

2. Pull the Stript Ease slowly toward you, keeping the leather down into the blade and flat against the Stript Ease. The Stript Ease can be adjusted to cut strips up to 2″ wide.

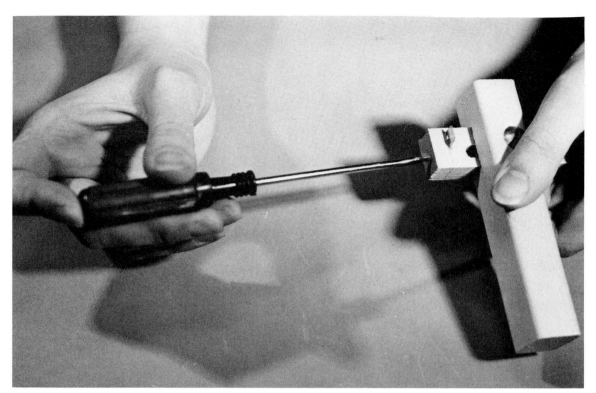

To change the blade in a Stript Ease, first loosen the screw that holds the old blade in place—you'll need a screwdriver with a small tip for this—then pull out the old blade, insert the new blade, and tighten the screw.

Injector blades can be used in the Stript Ease, but first break them in half with two pliers.

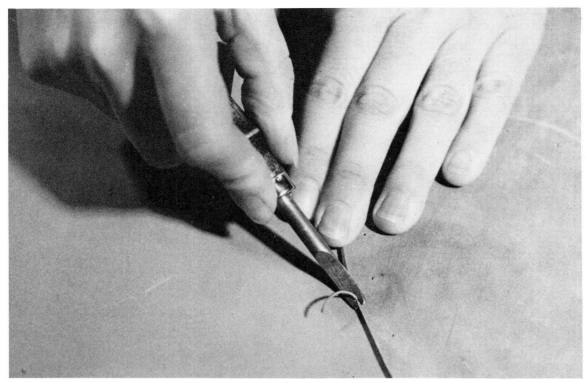

Beveling. To bevel the edges of the leather, hold your edge beveler as shown and push it against the top edge of the leather. A thin strip of leather will peel off.

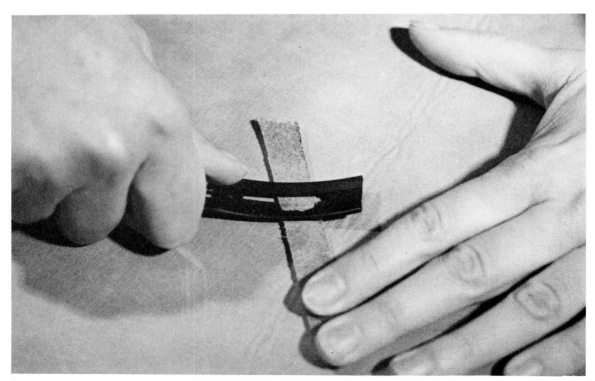

Skiving. To skive the end of a heavy piece of leather, place the leather on a piece of scrap so you won't cut your worktable. Then pull the skiving knife toward you as shown. Bear down lightly at first, going over the leather several times until you get the desired thickness.

Changing the Skiving Blade

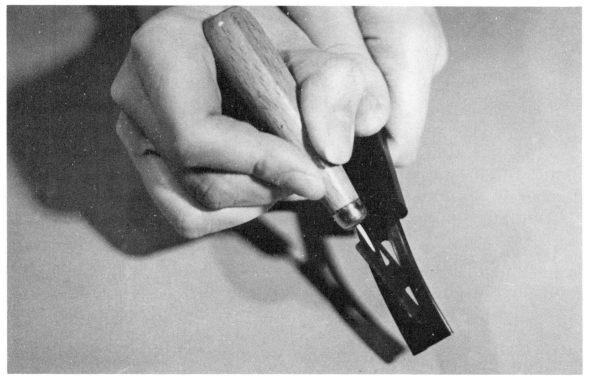

1. To change a worn blade in a skiving knife, use an awl to push out the old blade.

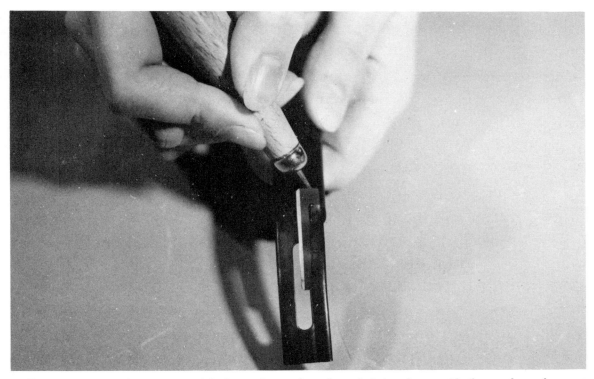

2. Then carefully place a new blade in the tool and push it in place with the awl as shown. Ordinary injector blades can be used in this tool.

Punching Holes

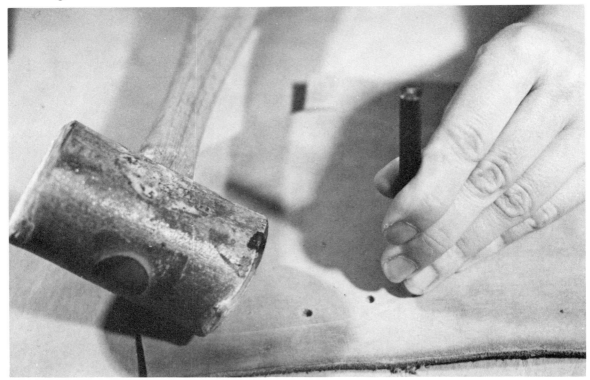

To punch round holes, place the leather on a steel plate or other hard surface. Put a piece of rubber or scrap leather under it to protect the hole punch. Hold the punch firmly and hit it squarely on top with a mallet.

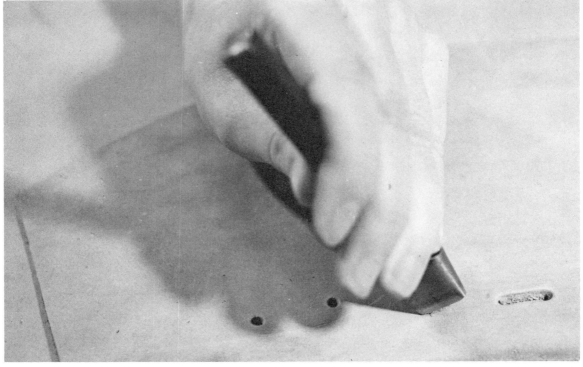

Use the same procedure to punch oblong holes. Since oblong holes are harder to punch, it will help if you rock the tool slightly forward, hit it with the mallet; then rock it slightly back and hit it again. Repeat this until the oblong hole has been cut cleanly through the leather.

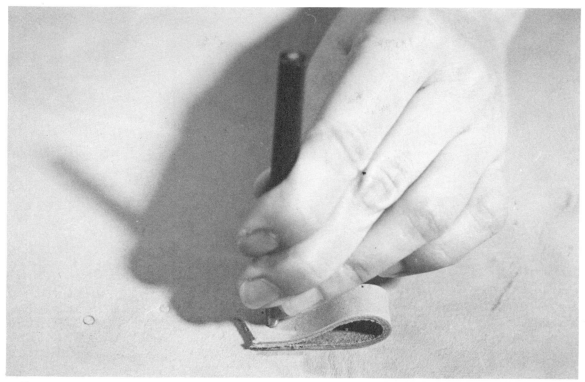

1. To set rivets, hold the two pieces of leather you're going to rivet tightly together and punch a hole with a #1 hole punch through both pieces.

2. Place a male rivet part up through the bottom hole.

3. Next, place a female rivet part over the top of the male rivet part.

4. Hold the rivet setter concave side down over the rivet and hit it squarely on top with a mallet. Make sure you do this directly on a steel plate or anvil otherwise the rivet won't set properly.

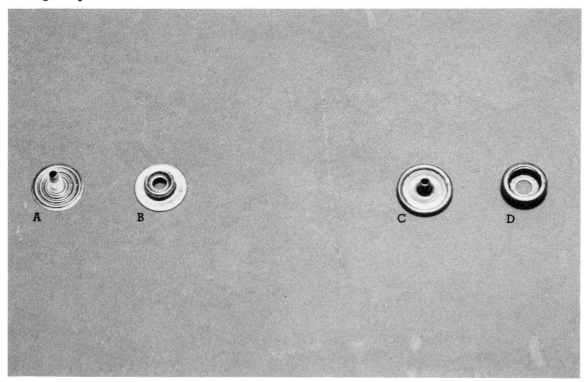

Here are all the pieces needed for one snap.

1. To set snaps (baby dot fasteners), punch a #2 hole in both pieces of leather. Place piece A up through one hole.

2. Place piece B over piece A.

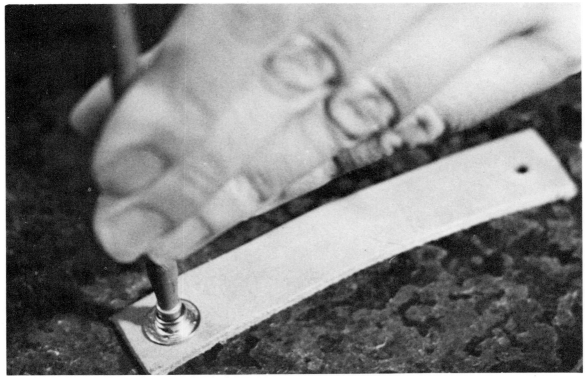

3. Holding the nub of the snap setter over the hole in piece A, hit it firmly with a mallet.

4. Turn the leather over and place piece C up through the second hole.

5. Place piece D over piece C and hit the snap setter firmly with a mallet to set.

6. When finished, the two halves should snap together as shown.

Dyeing

1. To dye latigo, first cover the edges of the leather with dye from a dauber.

2. Fold a piece of flannel or other absorbent material into a square to use for dyeing the surface of the latigo. Hold it over the bottle of dye and tip the bottle until you get some on the cloth.

3. Starting at the perimeter of the leather, gently rub in the dye using light, circular strokes.

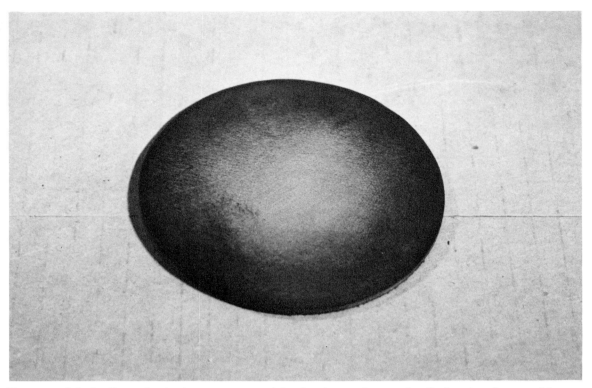

For a modeled effect, shade the perimeter of the leather only, gradually leaving less and less dye as you move toward the center of the leather.

For a solid, one-color effect, rub the cloth evenly over the whole piece of leather. You may have to dampen the cloth with dye several times.

4. When you're finished dyeing, the leather should be saddle soaped. This will protect the leather, bring out colors, and give a more finished look. Take a sheepskin scrap and rub it into a can of yellow saddle soap and then over the edge of the leather to smooth down rough fibers.

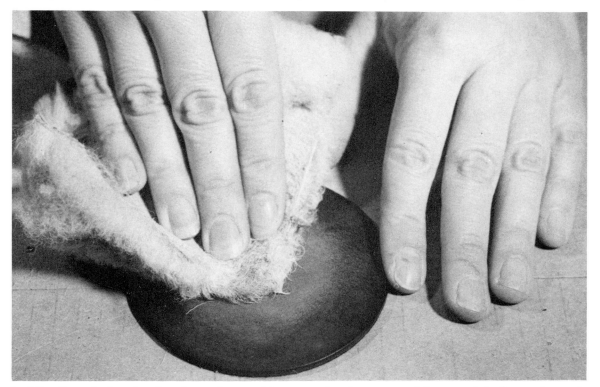

5. After the whole edge has been thoroughly soaped, rub more saddle soap into the surface of the latigo.

3 Barrettes

A barrette—a simple, useful, and popular item—is a good project for a beginning leathercrafter to start out with. With only a few tools, a small piece of leather, and your imagination, you can create a variety of barrettes. Large barrettes are worn by people with long, thick hair; mini-barrettes can be worn by almost anyone, and are especially popular with children.

Materials

For Large Barrettes:
1. Awl
2. Utility knife
3. Edge beveler
4. #10 hole punch (#3 hole punch for mini-barrettes)
5. Rawhide mallet
6. ¼" dowel (⅛" dowel for mini-barrettes)
7. Hack or coping saw (use metal shears for mini-barrettes)
8. Pencil sharpener
9. Dye and saddle soap
10. Sandpaper (for mini-barrettes)

Use leather that is at least 5/6 oz. and fairly stiff. See the end of this chapter for barrette patterns.

Problems

If the barrette doesn't hold its shape and the stick falls out easily, you probably used latigo that's too thin or too soft.

If the leather tears around the holes, you've probably made the holes too close to the edge of the leather.

Making a Large Barrette

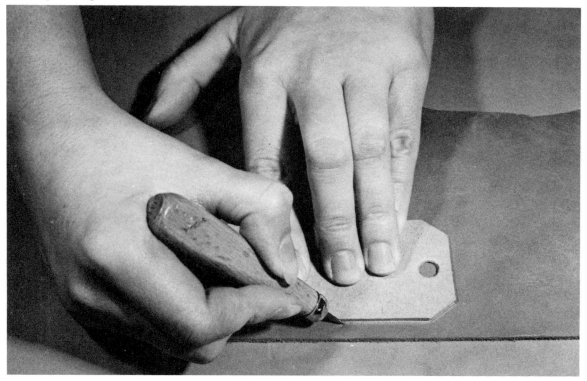

1. With an awl, trace a cardboard barrette pattern, including the holes, on a piece of latigo.

2. Carefully cut out the barrette with a utility knife.

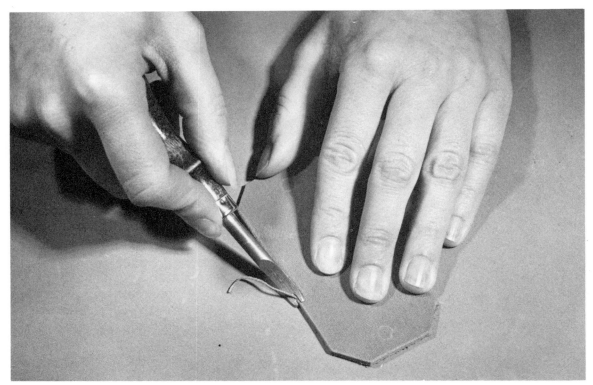

3. Using an edge beveler, bevel the edges of the leather.

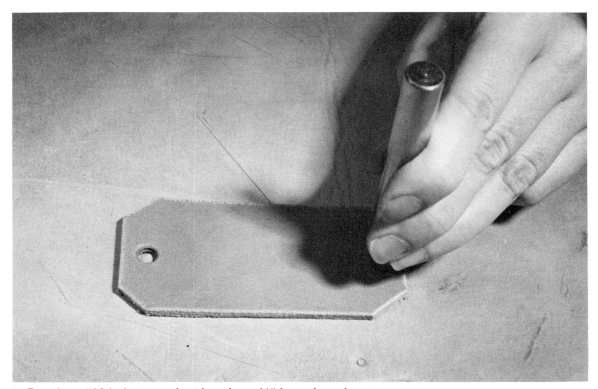

4. Punch a #10 hole in each side, about ¼" from the edge.

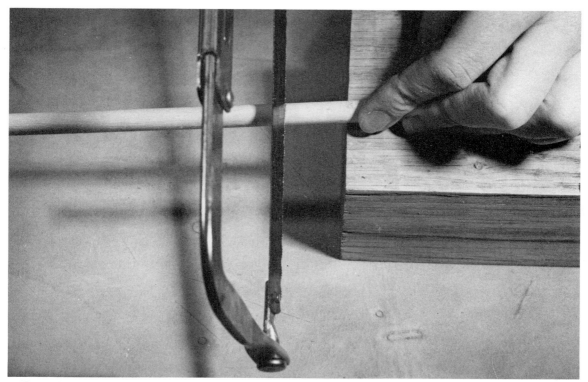

5. To make the stick, cut a piece of ¼″ dowel with a saw about 1½″ to 2″ longer than the barrette.

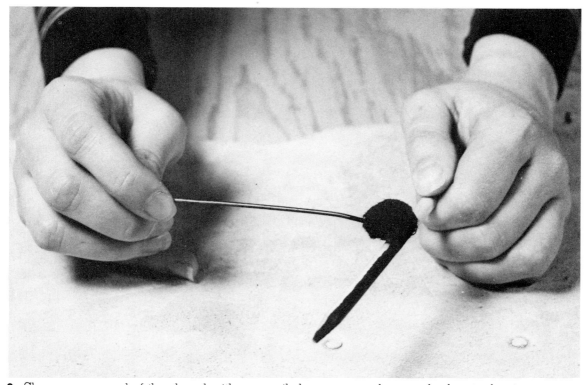

6. Sharpen one end of the dowel with a pencil sharpener and use a dauber to dye it.

7. Then rub the stick with saddle soap.

8. Insert the stick through the holes in the barrette and it's finished.

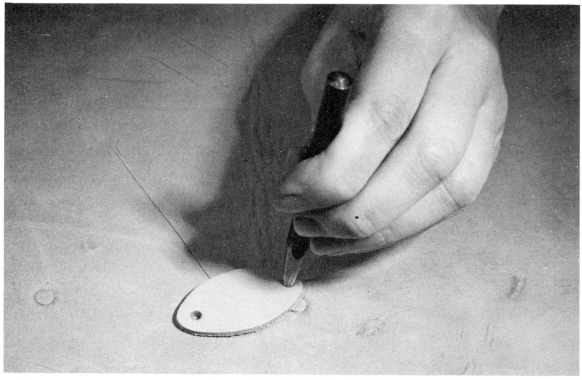

1. To make a mini-barrette, follow steps 1–3 for making the large barrette. Then use a #3 hole punch for the holes.

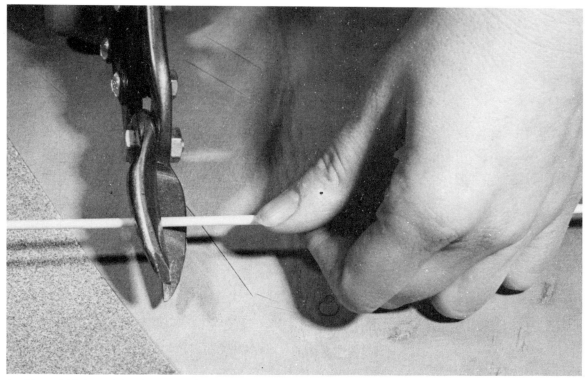

2. Use a ⅛″ dowel for the stick. Snip the dowel with a shears and smooth the ends to a point on a piece of sandpaper.

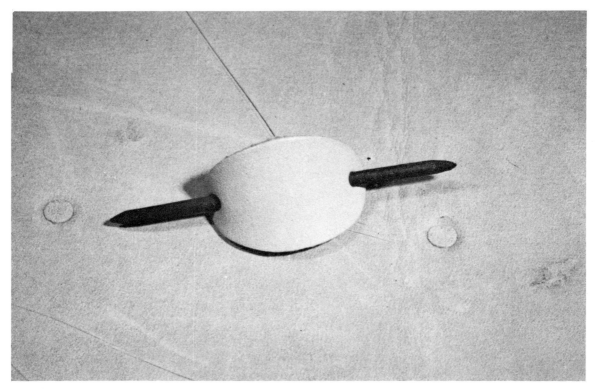

3. After dyeing and saddle soaping the stick, insert it in the holes of the barrette.

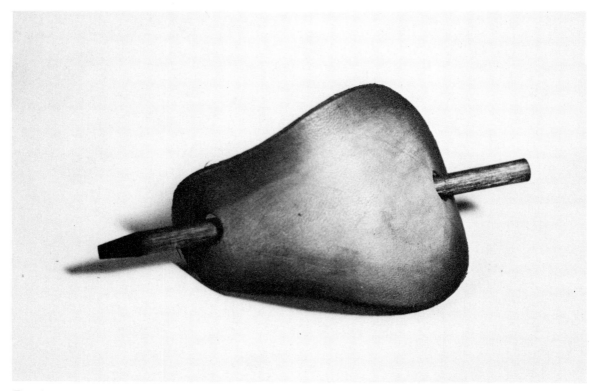

This barrette was cut in a free-form shape and shaded with dye. By Holy Cow Leather.

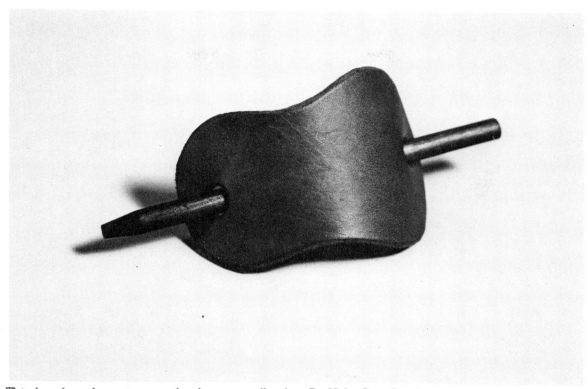

This free-form barrette was dyed an overall color. By Holy Cow Leather.

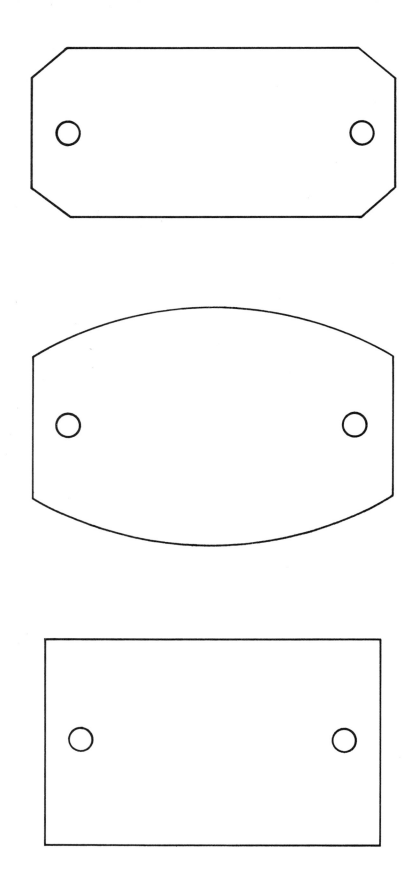

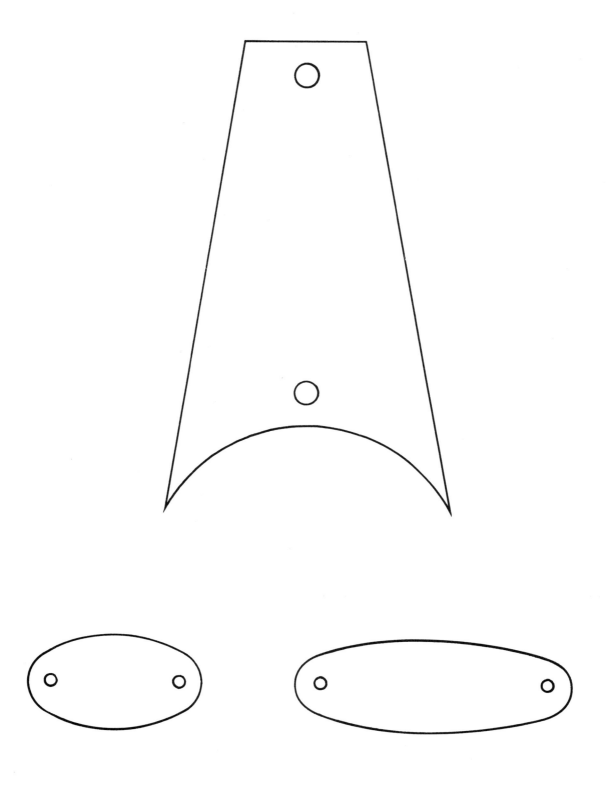

These are patterns for large barrettes and mini-barrettes.
Cut one piece for each barrette.

4 Keyrings

Another project that's easy to make and takes only small pieces or scraps of leather is a keyring. Keyrings can be made in a variety of shapes, sizes, and colors. Leather keyrings are easy to find when you're fumbling inside your pocket or handbag. You might try making a different shaped keyring for each key to help you find them in a hurry.

Materials

1. Awl
2. Utility knife
3. Edge beveler
4. Dye and saddle soap (optional)
5. #1 hole punch
6. Rawhide mallet
7. 1¼" metal keyring

8. Small rivets
9. Rivet setter

Latigo between 5/6 oz. and 7/8 oz. will work best for this project. See the end of this chapter for keyring patterns.

Problems

If the rivet goes in crooked, you may not have punched the hole straight through both pieces of leather. Also, be sure to leave enough room for the metal keyring to fit between the folded leather.

If the rivet comes out, you may not have hit the rivet setter hard enough. Or you may have left a piece of scrap leather under the keyring while riveting. Remember that riveting must be done on a hard, smooth surface such as a steel plate or anvil.

1. After tracing the keyring pattern and cutting it out in latigo, bevel the edges with an edge beveler. At this point, dye or otherwise decorate the piece of leather if you wish. If you dye the leather, it will look better dyed on both sides.

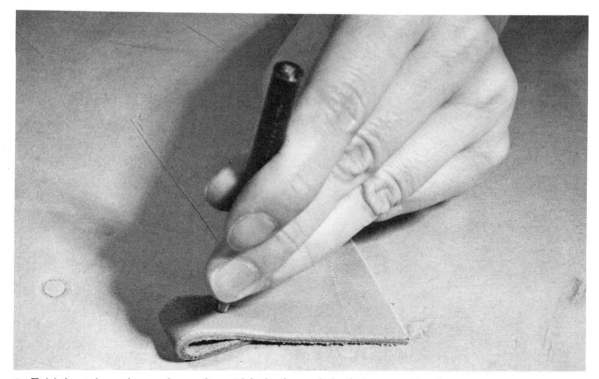

2. Fold the tab under and punch a #1 hole through both layers of leather.

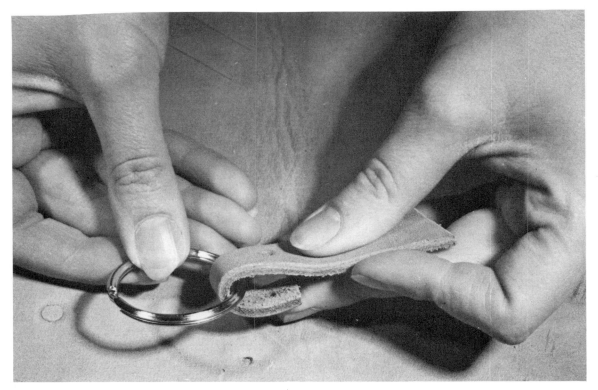

3. Place the metal keyring inside the folded tab.

4. Rivet the metal keyring in place with a small rivet. If you use a nickel keyring, use a nickel rivet.

This keyring was decorated with a leather-burner and dye. By Lyn Taetzsch.

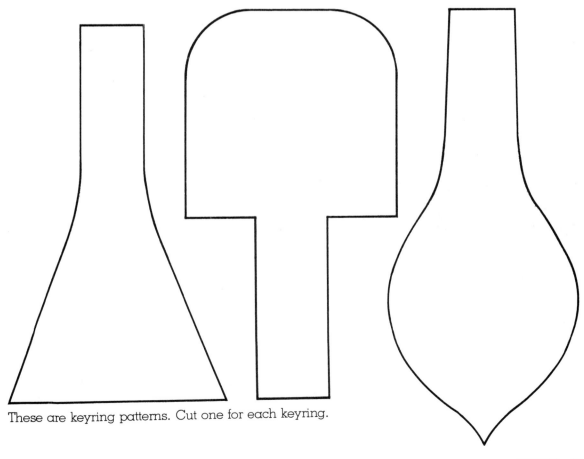

These are keyring patterns. Cut one for each keyring.

5 Earrings

Earrings can be made with earwires for pierced ears or with screw-on type findings for nonpierced ears. You can turn latigo scraps into stylish earrings—such as a long, dangling series of leather pieces—by using your imagination. Make them small and dainty, large and flashy, or whatever style suits you.

Materials

1. Awl
2. Utility knife
3. Edge beveler
4. #0 hole punch
5. Rawhide mallet
6. Dye and saddle soap (optional)
7. ⅛" or ¼" jumprings
8. 2 pairs of pliers (jewelry pliers work best)

9. Earwires

Small pieces of latigo no heavier than 7/8 oz. are best for earrings. Thicker leather can be used, although larger jumprings may be necessary. See the end of this chapter for earring patterns.

Problems

Be careful with nonsymmetrical shapes; turn your pattern over after cutting the piece for one earring so they'll hang in the right direction when worn. If the earrings dangle unevenly, it may be because you punched the holes unevenly.

1. Using your pattern, trace and cut two matching earring shapes from your latigo. Then bevel the edges and punch a #0 hole at the top of each shape. Dye or decorate the pieces of leather if you wish.

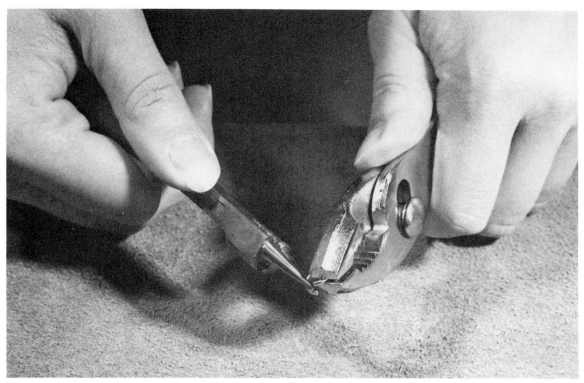

2. Spread open a ⅛″ or ¼″ jumpring as shown.

3. Put the jumpring through the hole in each leather shape and twist it closed with the pliers.

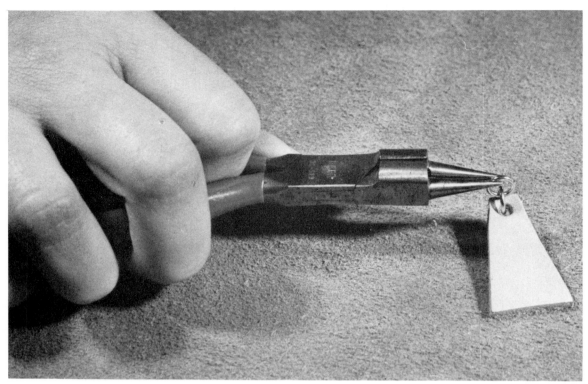

4. Using the same procedure, place another jumpring through the first one and twist it closed.

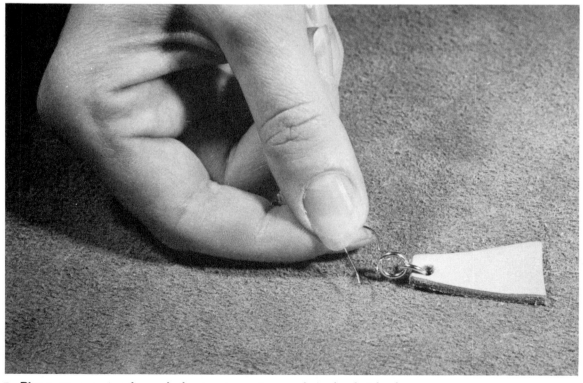

5. Place an earwire through the top jumpring and you're finished.

For long dangle earrings, cut out 2 or 3 shapes for each earring and punch a hole at the top and bottom of all but the bottom piece (this needs only a top hole). Attach the pieces to each other with jumprings and then attach the earwire.

These are earring patterns. Cut two for each pair of earrings.

6 Wristbands

Wristbands can be wide or narrow, and can be left plain or heavily decorated. Depending on the width, coloring, and decorative effects used, they will be enjoyed by men, women, and children. The wristbands in this chapter are made with snaps so they fit snugly and can be taken off. However, one couple asked me to make them each a permanently riveted wristband which symbolized their love for each other—they may still be wearing them.

Materials

1. Stript Ease
2. Utility knife
3. Edge beveler
4. Skiving knife
5. #2 hole punch
6. Rawhide mallet
7. Dye and saddle soap (optional)
8. Medium snap (baby dot fastener)
9. Snap setter

The best thickness of latigo to use for wristbands is 5/6 oz. to 8/9 oz. Thin leather won't hold up as well as heavy leather.

Problems

If the snap won't stay set, you may not have skived the leather thin enough. If the wristband is weak around the snap, you may have skived too much.

If the wristband is too tight, allow more than 1" overlap when measuring. If the wristband is too loose, allow less than 1" overlap.

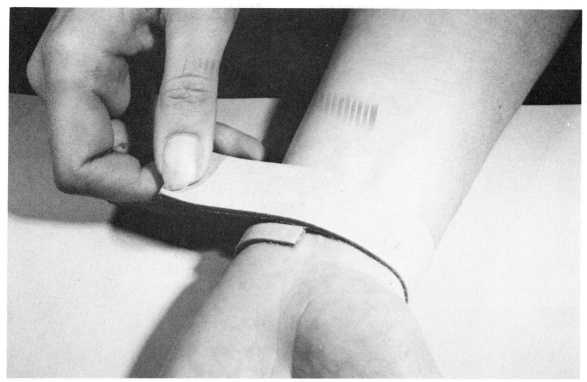

1. With your Stript Ease (see Chapter 2 for use), cut a strip of latigo from ½″ to 2″ wide and about 10″ long and wrap the strip around your wrist to judge the length. It should overlap about 1″.

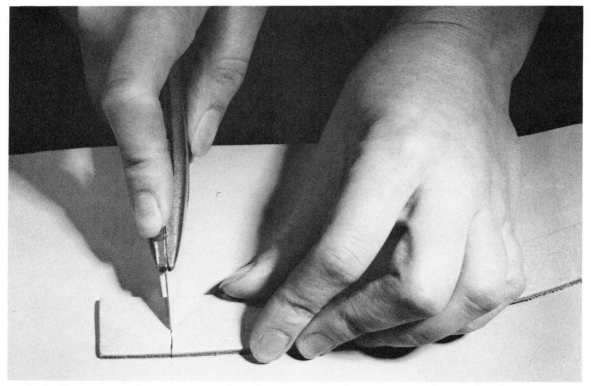

2. Cut off the excess length with a utility knife and bevel the edges with an edge beveler.

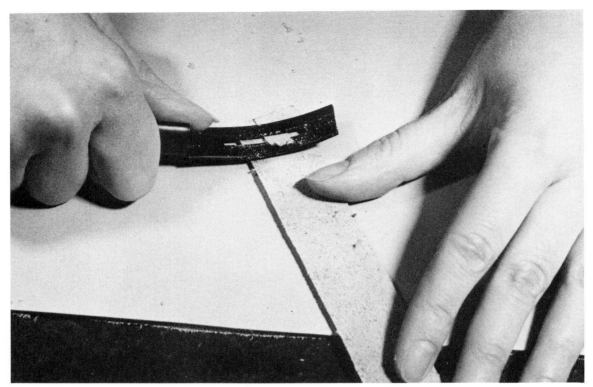

3. Skive with a skiving knife about 1″ in from each end of the strip. This will make it easier to put on the snaps, especially if you're using heavy leather.

4. Punch a #2 hole in each end of the strip, about ¼″ from each end. At this point, you may dye or decorate the leather.

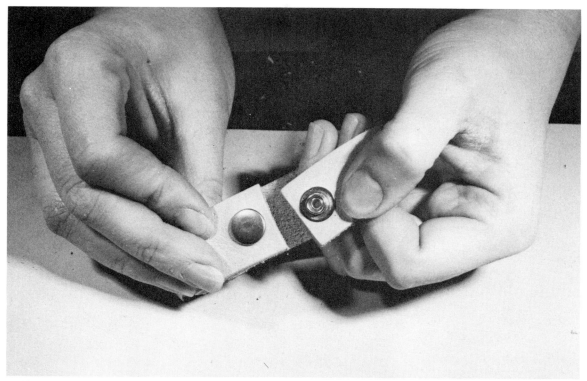

5. Place medium snap parts on each end, being careful to place them so they snap together correctly in an overlapping position the way you'll be wearing the wristband (refer to Chapter 2 for instructions on setting snaps).

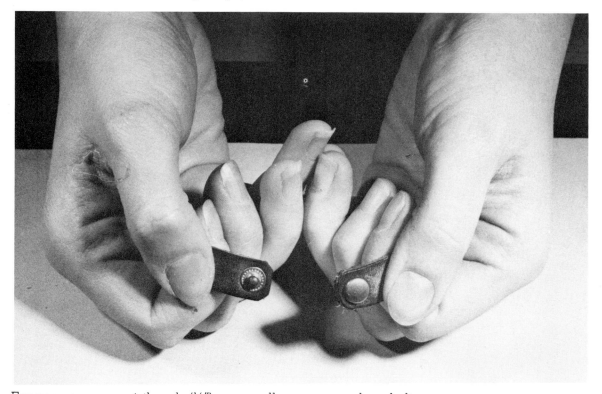

For very narrow wristbands (½″) use smaller snaps, such as belt snaps.

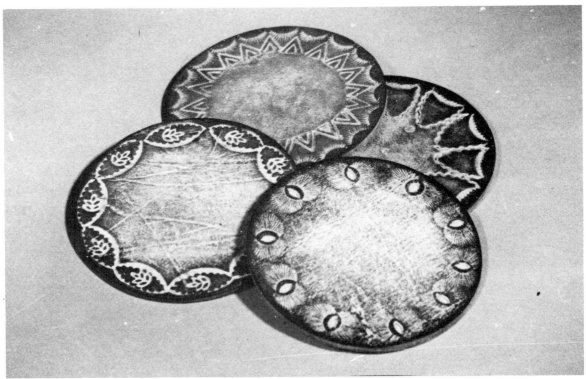

These coasters were made by cutting circles out of latigo scraps. By Holy Cow Leather.

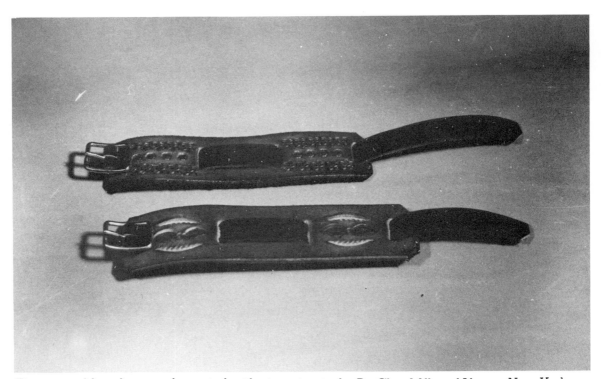

These watchbands were decorated with stamping tools. By Glen Miller of Ithaca, New York.

(Opposite page) This pencil holder was made by gluing pieces of latigo around a square of ½″ plywood. Strips of 2 to 2½ oz. kip leather were used to reinforce the corners. By Holy Cow Leather.

7 Watchbands

For a number of years, wide leather watchbands have been in fashion. Natural latigo makes an attractive wide band, with English Kip or another thin leather used for the strap that goes through the watch. These watchbands can be custom fit to your watch and your tastes. You can make several and change bands to match your mood or clothing.

Materials

1. Awl
2. Edge beveler
3. Oblong hole punch (½" for woman's watch, ⅝" for man's watch)
4. Rawhide mallet
5. Metal ruler
6. Utility knife
7. Dye and saddle soap (optional)
8. Bar buckle (½" for woman's watch, ⅝" for man's watch)
9. #0, #1 hole punches
10. Small rivet
11. Rivet setter
12. Scissors

You'll need some latigo, about 5/6 oz. to 8/9 oz., and English Kip or other thin but strong leather, about 2 oz. to 2½ oz. See the end of this chapter for the watchband pattern.

Note: Most men's watches will require a ⅝" strap and therefore a ⅝" buckle. Most women's watches vary from ¼" to ½". This kind of watchband looks best on a watch at least ⅜" wide, so a ½" buckle should work on most women's watches.

Problems

If your watch buckles up over the band, you've probably made the center oblong holes too close together.

If your watchband strap stretches a lot or breaks, the leather you're using is not strong enough for the strap. Be sure to use the 2 oz. to 2½ oz. English Kip or other similar leather.

If your strap doesn't fit smoothly through the oblong holes, use a larger oblong hole punch and repunch the holes.

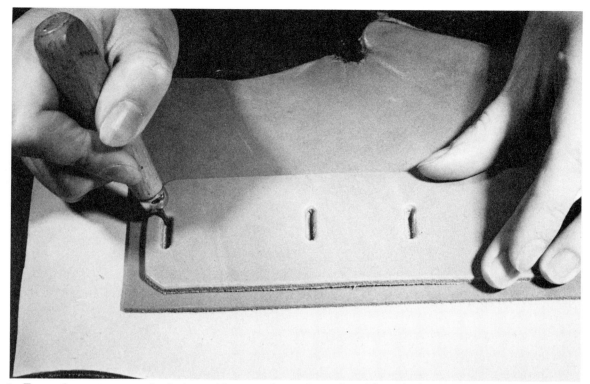

1. Trace the pattern on the latigo with an awl, marking the oblong holes so you know where to punch them. This band should be at least ½" wider than your watch, and just long enough to almost go around your wrist. After cutting the piece out, bevel the edges of the leather.

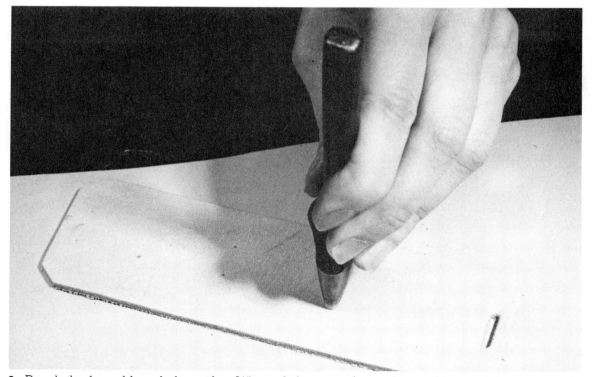

2. Punch the four oblong holes with a ⅝" punch for a man's watch, or a ½" punch for a woman's watch—one at each end and two in the center spaced about ¼" farther apart than your watch length. Dye or otherwise decorate the band if you wish.

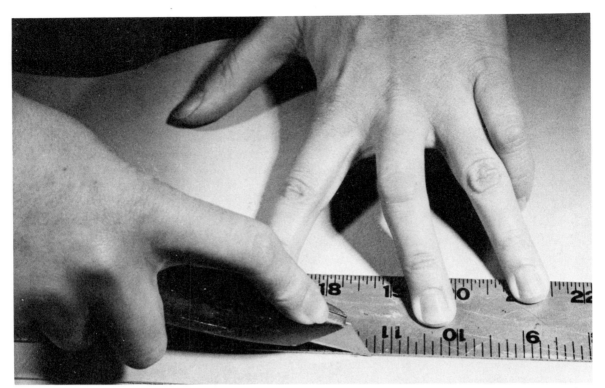

3. Using a metal ruler and utility knife, cut a 10" long strip of thin leather, ⅝" wide for a man's watch and ½" wide for a woman's watch. Measure the old strap on a woman's watch first—it may be less than ½". The strap you cut must fit through the watch pins.

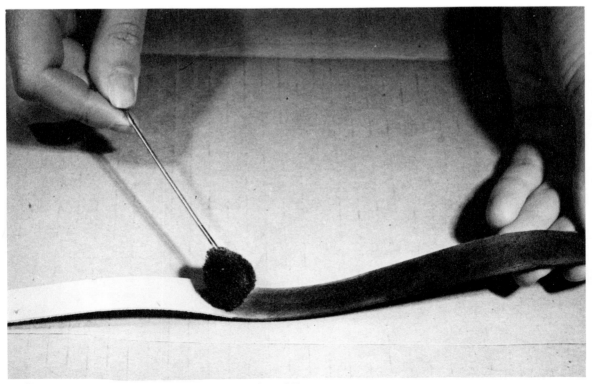

4. Dye the strap the color you want it and saddle soap it.

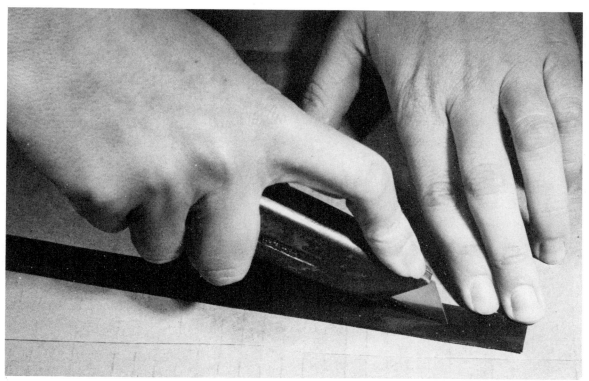

5. To put a buckle on the strap, first cut a ½″ slit about ¾″ from the end of the strap.

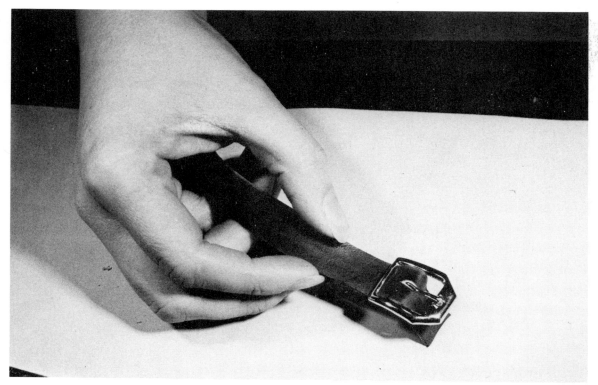

6. Push the tongue of the buckle up through the slit and fold back the remaining short end.

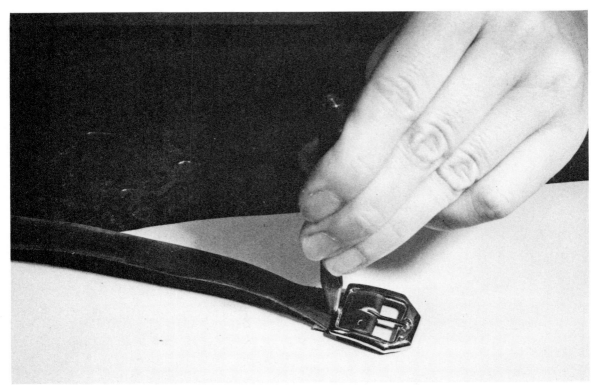

7. Punch a #1 hole just below the bar of the buckle, through both layers of leather.

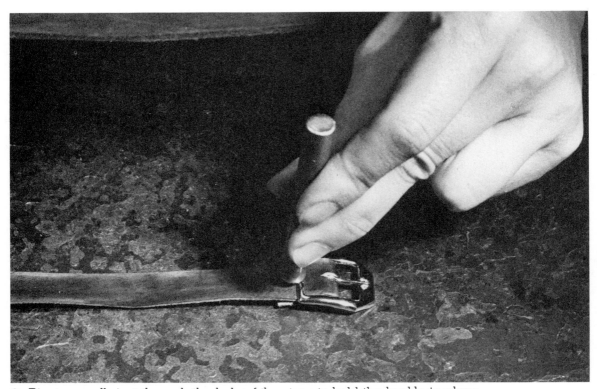

8. Rivet a small rivet through the hole of the strap to hold the buckle in place.

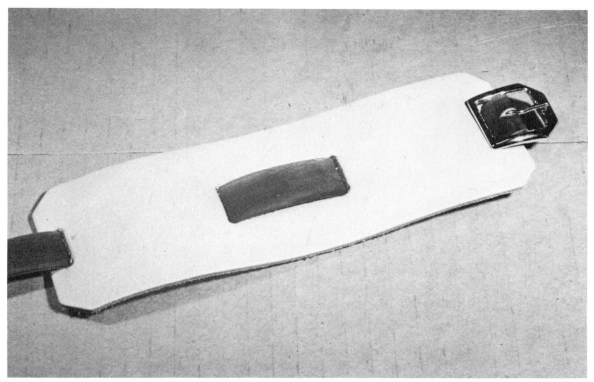

9. Weave the strap through the latigo band and slip your watch on the strap when you get to the center.

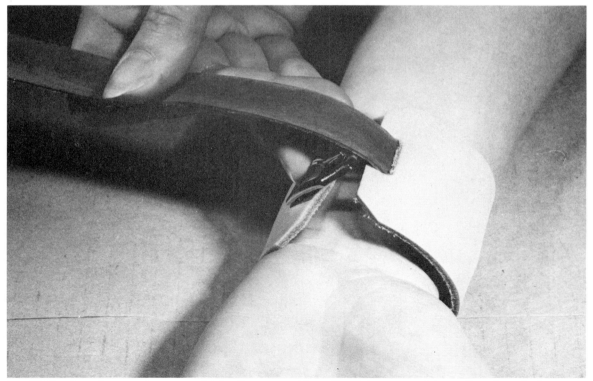

10. Fit the band around your wrist for the correct length. Put your watch on the strap before measuring or allow 1" extra if your watch is not on the strap.

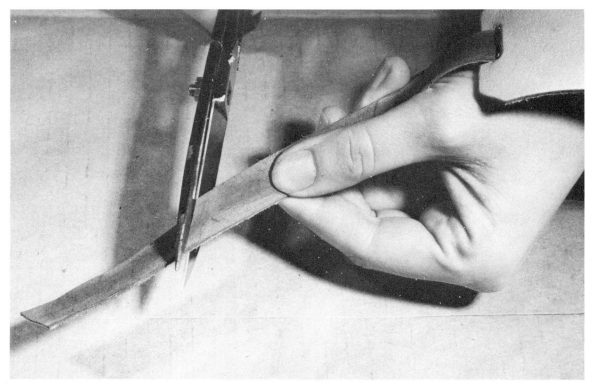

11. Cut off the excess strap with a scissors.

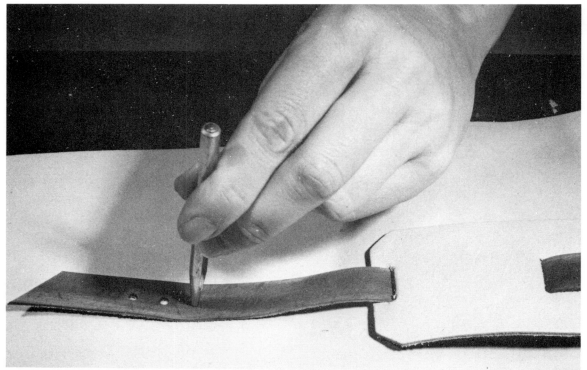

12. Punch two or three holes with a #0 hole punch.

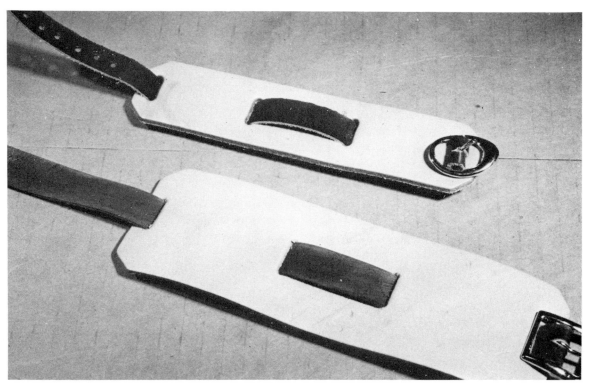

Your watchband is finished.

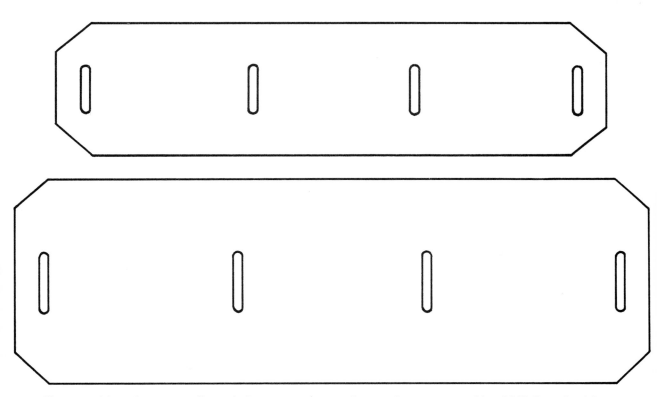

The watchband at top will work for women's watches with a strap width of ⅜". Punch oblong holes with a ½" punch. The watchband at bottom will work for men's watches. Punch oblong holes with a ⅝" punch.

8 Belts

Belts are one of the most popular items made from latigo leather. You can make traditional men's belts in the width of 1¼", high-fashion "skinnies of ½" and ¾" widths, and belts for jeans in the popular 1¾" width. They can be highly decorated or left plain to accent a striking buckle. You'll find that for yourself, your friends, and your family, the belts you make will be much in demand.

Materials

1. Long metal ruler
2. Stript Ease
3. Utility knife
4. Edge beveler
5. Dye and saddle soap (optional)
6. Skiving knife
7. 1" or 1½" oblong hole punch (½" or ⅝" for narrow belts)
8. Rawhide mallet
9. Buckle
10. #1 hole punch
11. Medium rivets (or small rivets for narrow belts)
12. Rivet setter
13. Awl
14. #7 hole punch (or #3 for narrow belts)

For belts 1¼" and wider, use a heavy latigo—8/9 oz. or 9/10 oz. For 1" belts or narrower, use a lighter weight—5/6 oz. or 7/8 oz.

If you use harness, or bar, buckles you'll need an oblong hole punch. Buckles that hook into the hole of a belt (and have no tongue) don't require an oblong punch.

Problems

It takes practice to be able to cut belts properly with a Stript Ease. Here are some helpful hints: always keep the blade sharp and in tightly. Give yourself plenty of room to work and always start with a perfectly straight edge on the right side of the leather. You'll also find that some kinds of leather are easier to cut than others. Leather that is very stiff is difficult to cut, but so is leather that is very soft.

If your buckle ends up off-center or the back piece of the strap holding the buckle doesn't match the front of the belt, you probably haven't punched the oblong hole properly. Be sure this hole is in the center and parallel to the sides.

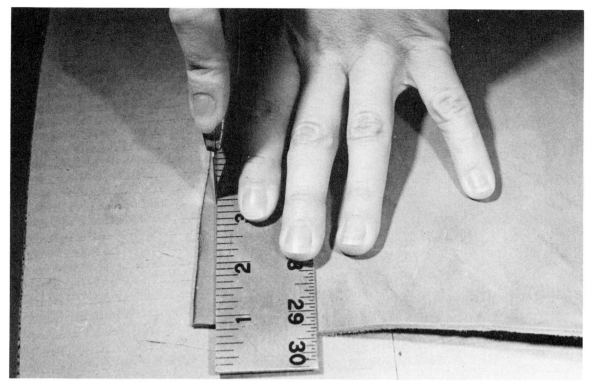

1. Lay the latigo out on a board or layers of cardboard to cut. Give yourself plenty of room. Make sure the leather is long enough for the belt size you need. Using the longest ruler you have, cut a straight edge on the right side of the leather.

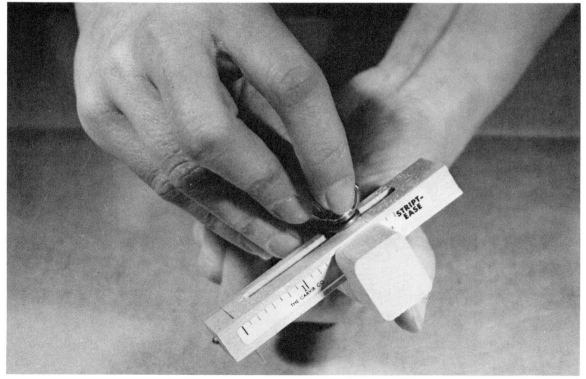

2. Now, use your Stript Ease (as shown in Chapter 2) to cut a belt strip. Make sure you adjust the width first by loosening the nut in order to slide the wood pieces to the proper setting.

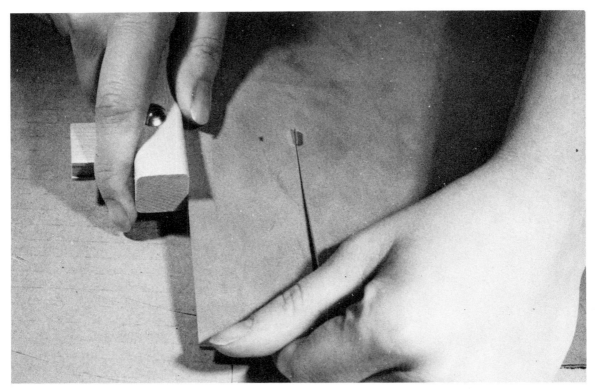

3. In this demonstration we're making a 1¾" belt. Note how the left hand holds the leather steady while the right hand controls the Stript Ease.

4. Allow an extra 8" beyond waist size (or hip size) and cut off the end of the belt strip with a utility knife.

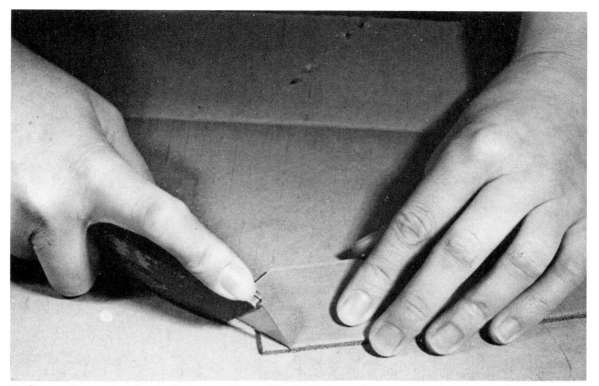

5. Shape the end of the belt strap with a utility knife to make a rounded or pointed tongue.

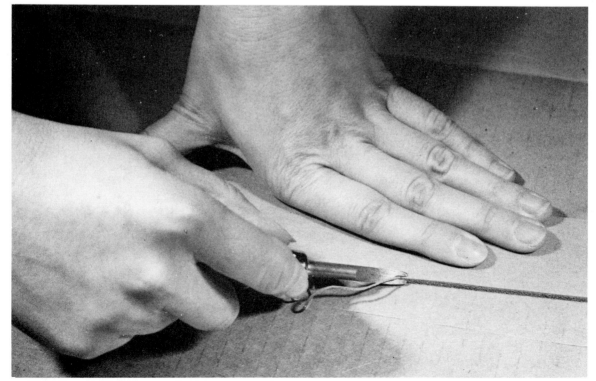

6. Then bevel the edges of the belt strap with an edge beveler and dye or decorate the leather if you wish.

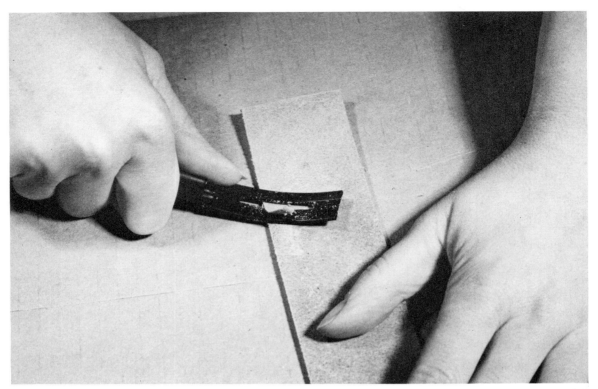

7. If you're using heavy leather—8/9 oz. or more—skive about 2″ in from the end and about ½ the thickness of the leather. This will make it easier to put the buckle on.

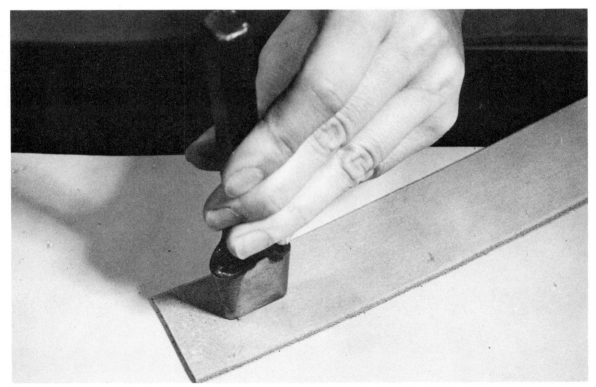

8. Punch an oblong hole about 1″ from the skived end of the belt as shown. This hole must be punched in the center, parallel to both sides of the belt.

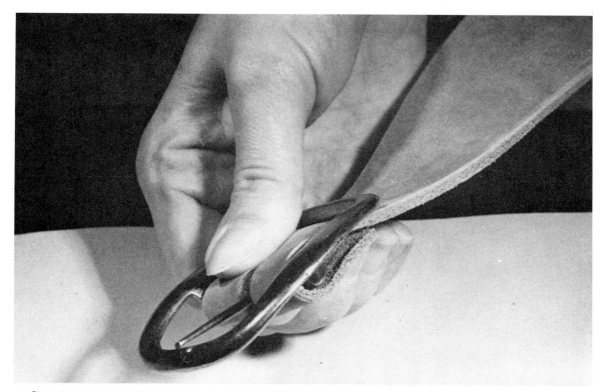

9. Slip the belt strap through the buckle, pushing the tongue of the buckle up through the oblong hole.

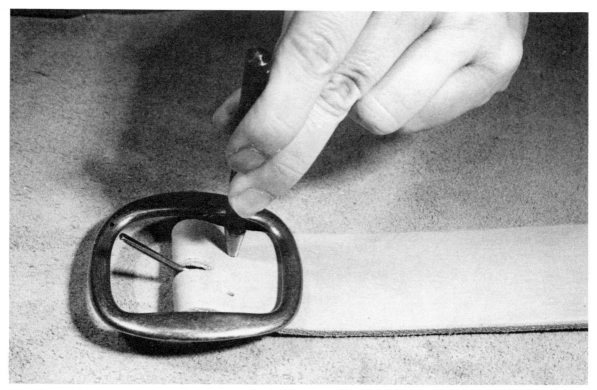

10. Then fold the end of the belt back evenly so the halves of the oblong hole match and the sides of the leather match.

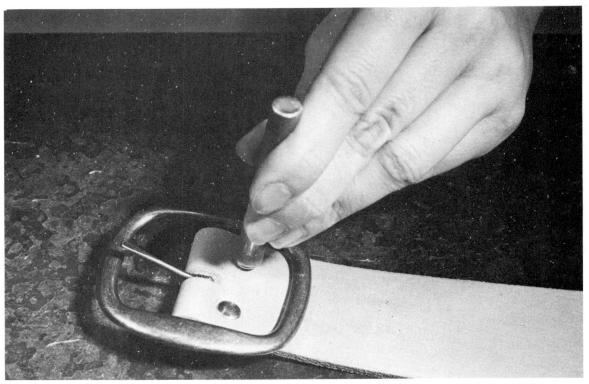

11. Holding the leather in place, punch a #1 hole on each side of the tongue, about ¼" below the center bar and through both layers of leather.

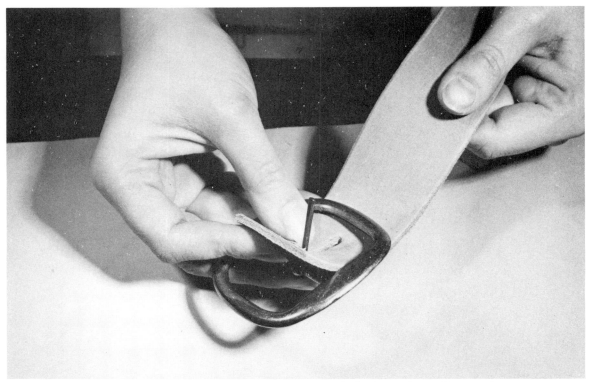

12. Rivet the buckle in place with medium rivets for wide belts and small rivets for narrower belts. A #1 hole is sufficient for both medium and small rivets.

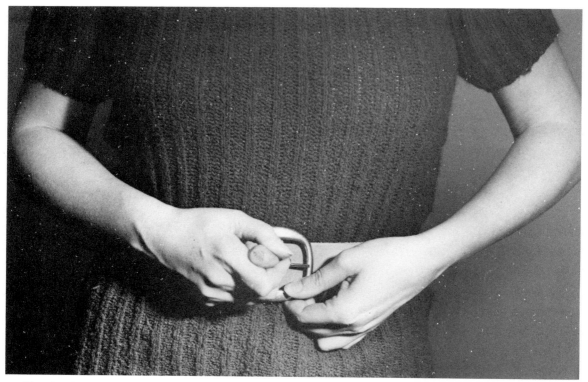

13. Put the belt around your waist and with an awl mark where you want to punch the holes.

14. Punch the holes with a #7 hole punch for wide belts and a #3 hole punch for narrow belts.

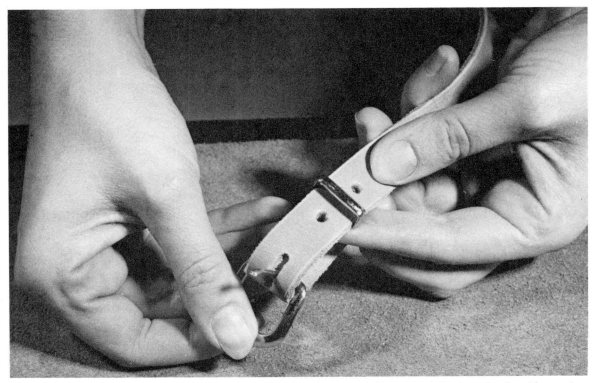

Harness buckles require a keeper, which should be slipped on the belt before the buckle and riveted in place.

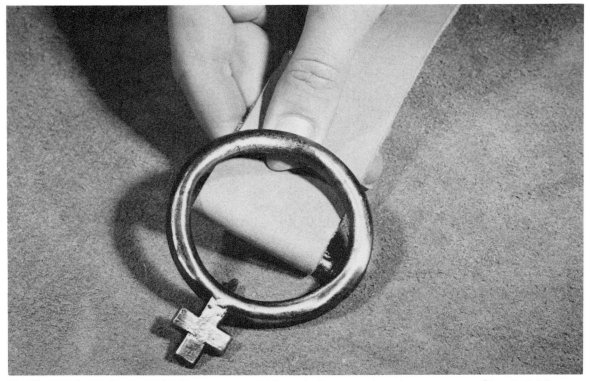

Buckles such as this one don't require an oblong hole since there's no tongue. Simply fold over the end of the belt and rivet in place.

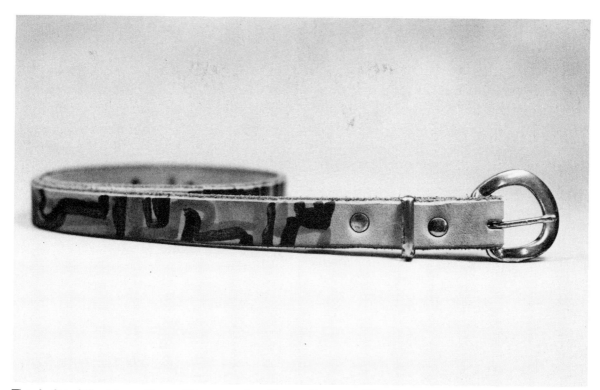

This belt is ¾″ wide, with a gilt harness buckle and keeper. The design was made with dye and paintbrush. By Lyn Taetzsch.

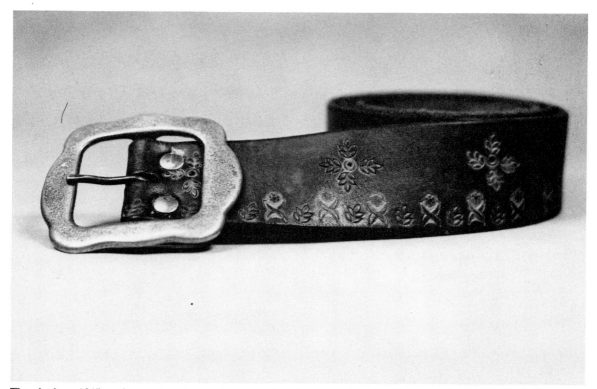

This belt is 1¾″ wide, with a brass buckle. By Mary Azerbegi of Braum's Leather, Berkeley, California.

9 Belt Pouch

Belt pouches are handy for carrying small items while hiking, camping, or even during your daily routine because they attach to your belt and leave your hands free. Making a belt pouch of sturdy latigo will protect the items you carry from damage. The one shown in this chapter will fit on belts up to 3″ wide.

Materials

1. Awl
2. Utility knife
3. Edge beveler
4. #1, #3 hole punches
5. Rawhide mallet
6. Dye and saddle soap (optional)
7. Medium rivets
8. Rivet setter
9. Medium snaps
10. Snap setter
11. 36″ length of ⅛″ latigo lace
12. Scissors
13. Latigo life-eye lacing needle

Latigo that is 6/7 oz. to 7/8 oz. will work best for belt pouches. A belt pouch pattern is at the end of this chapter.

Problems

If it's too difficult to pull the lace through the holes, use a #4 hole punch instead of a #3 on the lacing holes.

If you can see lace showing between the main piece and sides of the pouch, you probably haven't laced it tightly enough. We can't stress strongly enough the importance of lacing a bag tightly. If necessary, use a pliers to help pull the lace tight. Always pull on the lace, never on the needle. If the lace should break, tie a knot inside the bag close to the leather and start lacing again in the next hole. Tie the last knot firmly against the leather so the lace can't loosen up.

Always keep an extra lacing needle on hand because occasionally a piece of lace will get stuck in the needle that you won't be able to get out.

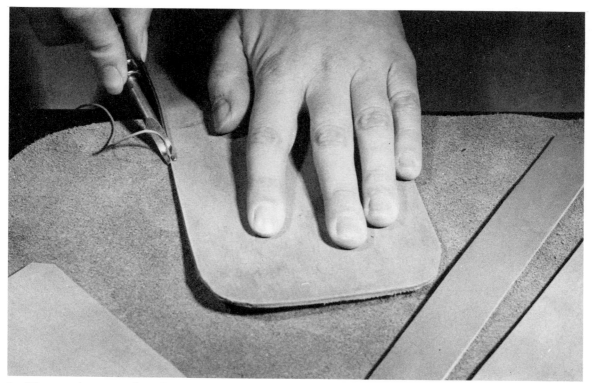

1. After you've traced and cut out all pieces, bevel the edges with an edge beveler.

2. Mark the lacing holes, the snap holes, and the rivet holes with an awl. Then punch the lacing holes with a #3 hole punch and the snap and rivet holes with a #1 hole punch. You may dye or decorate the pieces if you wish.

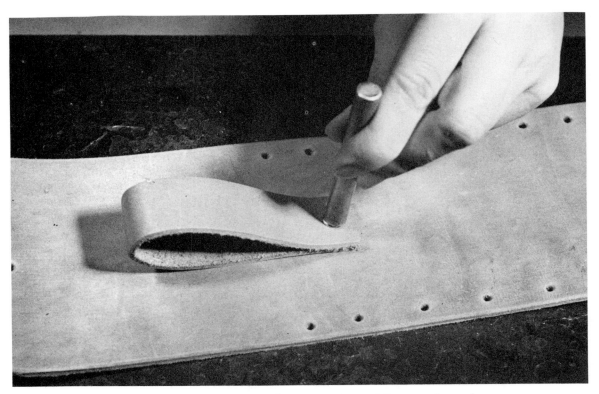

3. Next, fold the strap in half and rivet it to the main piece of the pouch as shown.

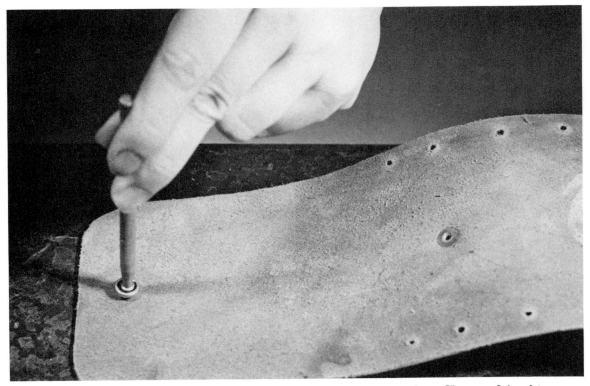

4. Insert snap pieces where marked and set them with snap setter (see Chapter 2 for this procedure).

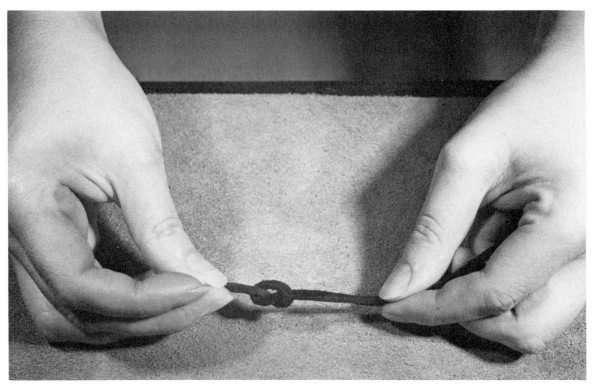

5. Knot one end of a 36″ length of ⅛″ latigo lace.

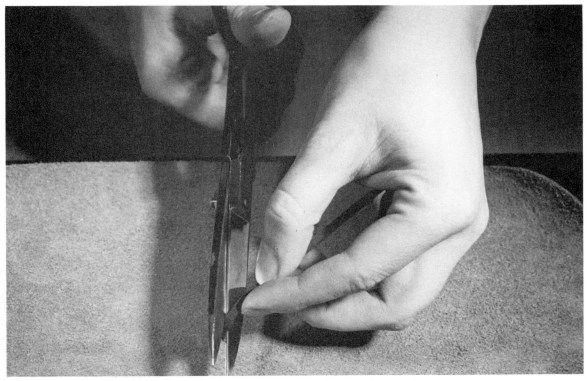

6. With a scissors, cut the other end of the lace at a diagonal to make a point.

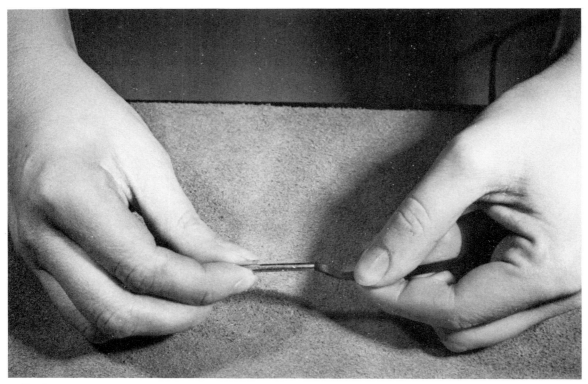

7. To thread on a latigo life-eye needle, screw the needle onto the latigo point by turning it in a clockwise direction.

8. Begin lacing by pulling the needle from the inside top of one side, as shown, and through the top hole on the main piece. Be sure the pouch front and back overlaps the sides.

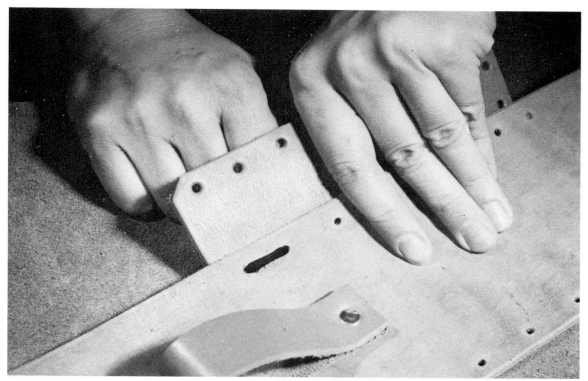

9. Push the needle down through the second hole in the main piece and the side. Be sure to pull tightly.

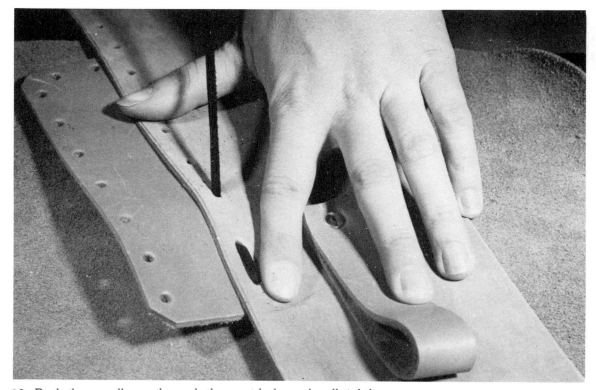

10. Push the needle up through the next hole and pull tightly.

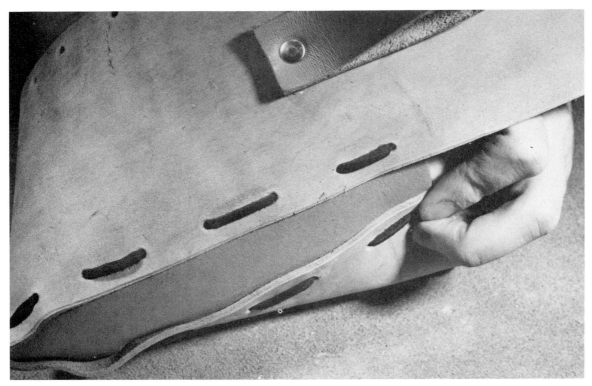

11. Continue until the side is completely laced.

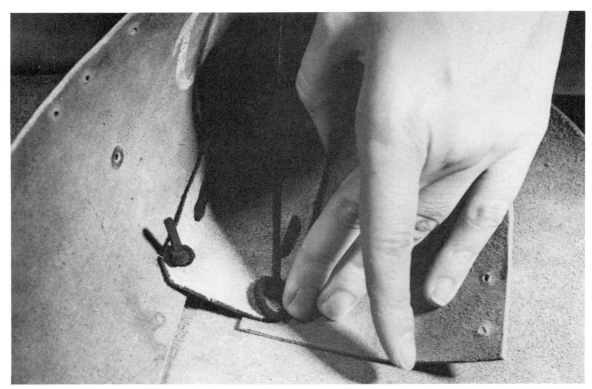

12. When you reach the end of the side, tie a single knot tightly against the leather and cut off the excess lace.

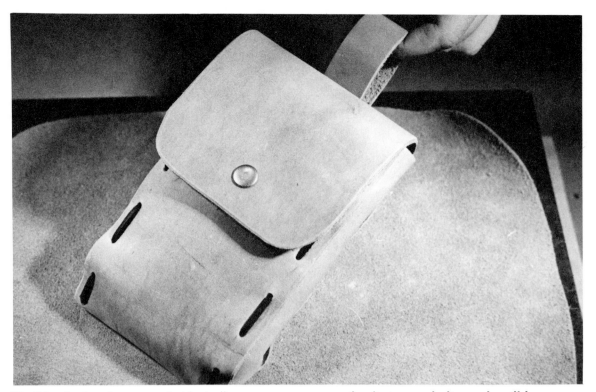

13. After lacing the second side in the same manner as the first, your belt pouch will be completed.

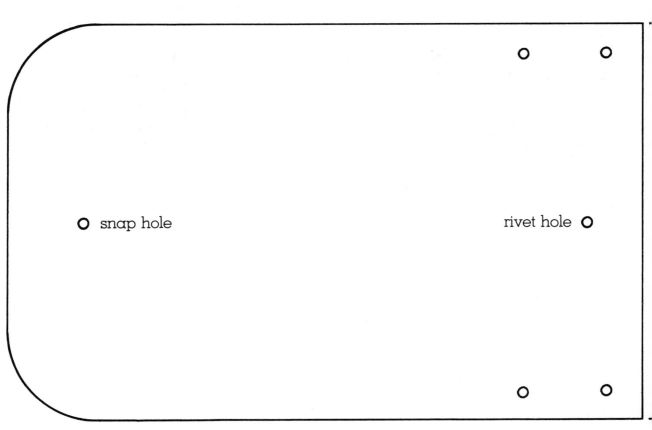

The distance shown by the broken lines should measure 4⅜″ and this belt pouch pattern will be life size. Cut one piece.

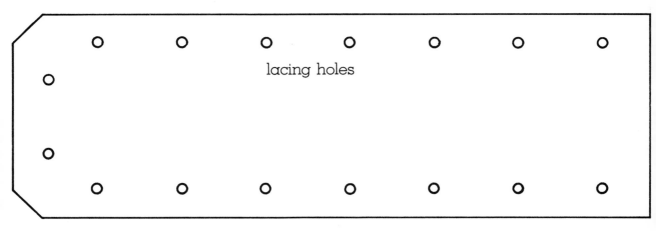

Cut two pieces of this pattern of a belt pouch side.

lacing holes

snap hole

rivet hole

4⅜"

rivet hole

Cut one piece of
this pattern of a belt
pouch strap.

10 Three-Legged Stool

The stool shown in this chapter is simple in design, easy to construct, and requires few tools to make. It can be taken apart and folded up for storage or carrying. With shorter legs, this stool is ideal for a child.

Materials

1. Three 1″ dowels (each 24″ long)
2. Sandpaper
3. Drill
4. Three-pronged ¼″ bolt (each prong 1¼″ long)
5. Three ¼″ nuts
6. Awl
7. Utility knife
8. Edge beveler
9. #1 hole punch
10. Rawhide mallet
11. Dye and saddle soap (optional)
12. Rivets
13. Rivet setter

Latigo 7/8 oz. or heavier works best for the seat part of this three-pronged stool. Use the pattern at the end of this chapter. The bolt and predrilled legs can be purchased from Tandy Leather Co. (see Suppliers List for address).

Problems

If the rivets are set crooked, you may not be punching the holes straight through both pieces of leather. Be sure to mark and punch the holes carefully and hit your rivet setter directly over each rivet. Remember that all riveting should be done on a piece of steel or marble, or on an anvil.

If the stool doesn't balance properly, you may have drilled the bolt holes at uneven distances from the bottoms of the legs. Check this and adjust the length of the legs accordingly.

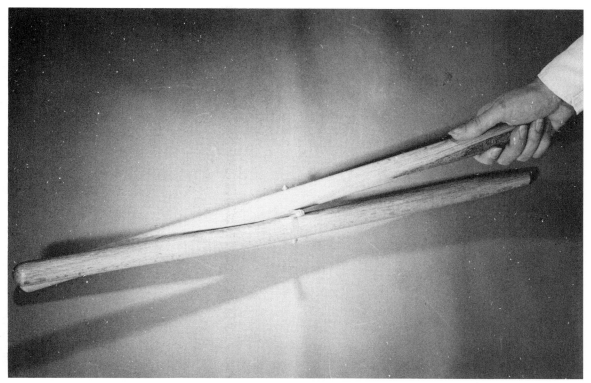

1. Sand the bottom of three 1″ dowels round and sand the tops flat and smooth. Then shellac or finish the wood as desired. Drill a ⅜″ hole 11½″ from the bottom (round end) of each leg or use predrilled legs from Tandy Leather. Insert the three-pronged bolt through the hole in each leg.

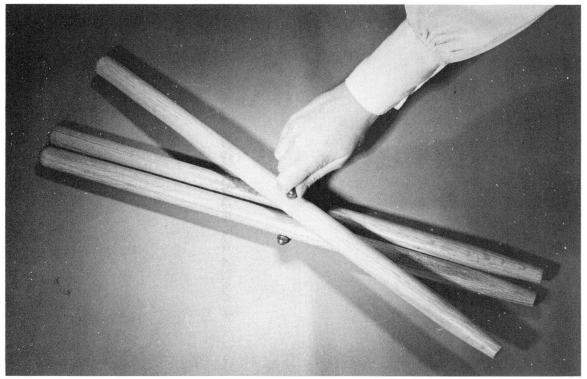

2. Screw a nut on the end of each prong.

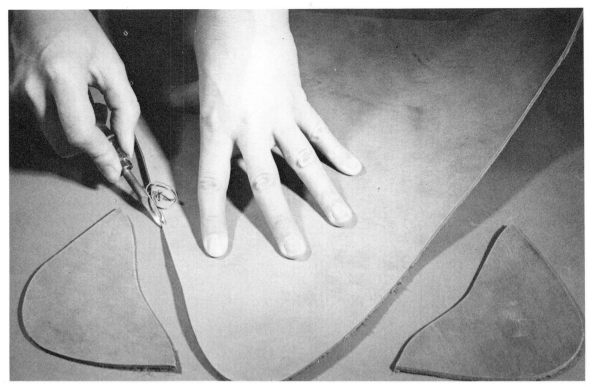

3. With an awl, trace the seat and corner pocket pieces and cut them out with a utility knife. Bevel the edges with an edge beveler.

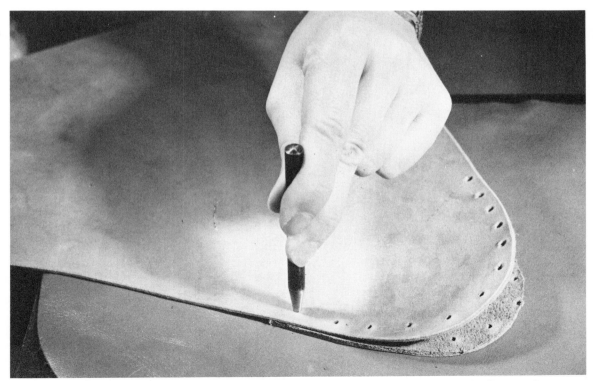

4. Next, mark the holes with an awl and punch them out with a #1 hole punch. Dye or decorate the leather if you wish. Be sure to punch through the seat and pocket pieces at the same time so the holes will match perfectly. Keep the flesh sides of the leather together.

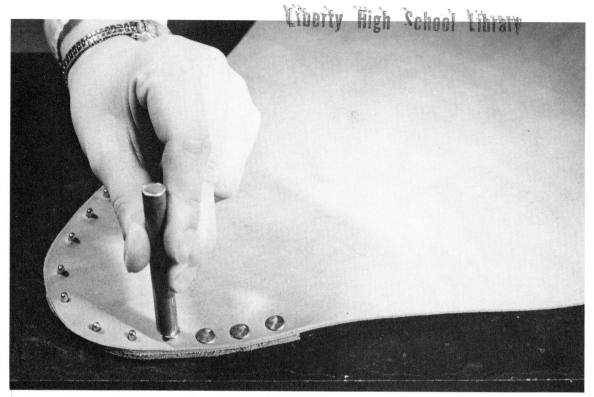

5. Rivet with a rivet setter each corner pocket piece to the seat as shown.

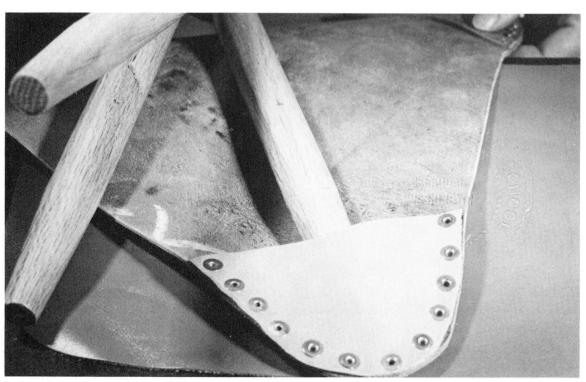

6. Insert the flat tops of the legs into the seat pockets.

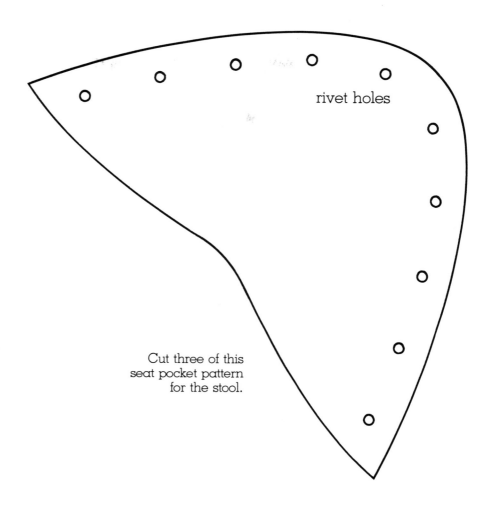

rivet holes

Cut three of this
seat pocket pattern
for the stool.

Your three-legged stool is completed.

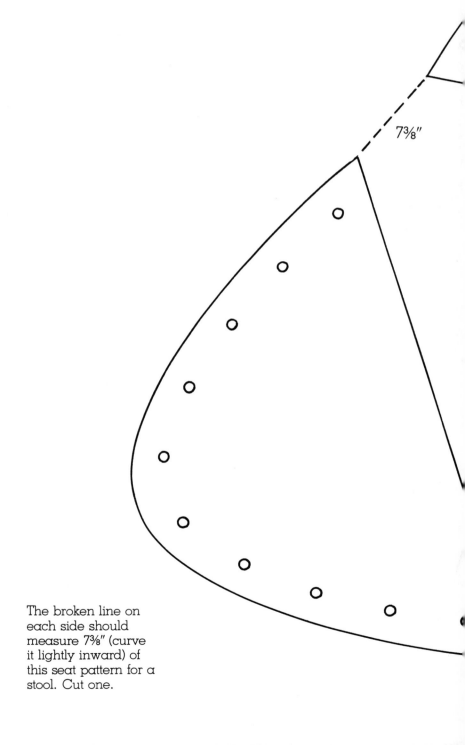

7⅜″

The broken line on
each side should
measure 7⅜″ (curve
it lightly inward) of
this seat pattern for a
stool. Cut one.

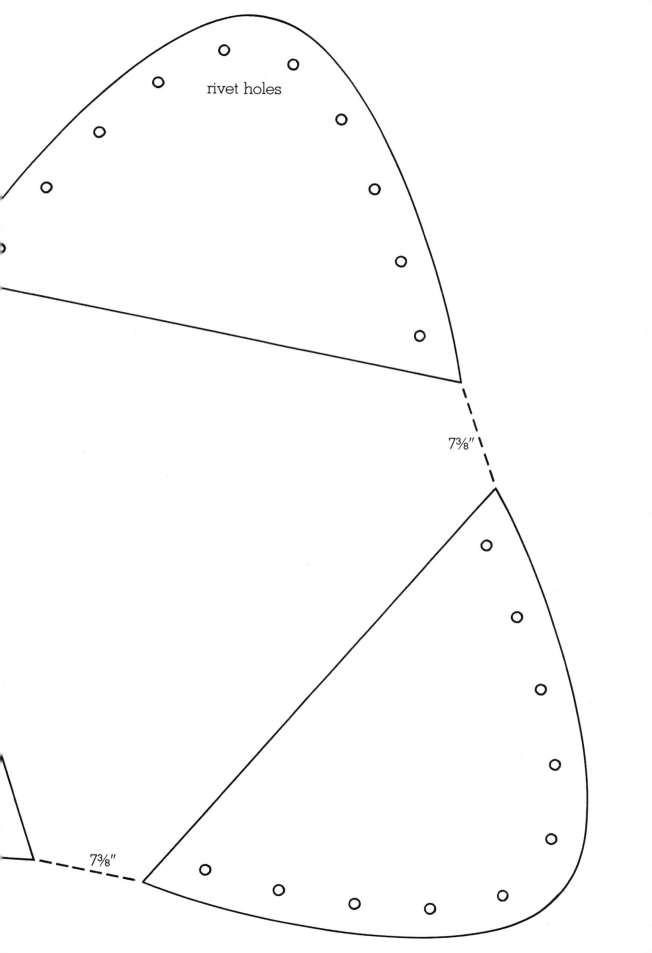

rivet holes

7⅜"

7⅜"

11 Sandals

Leather sandals have been used for centuries and are still popular today. They're attractive, comfortable, and never seem to wear out. We're going to show you how to make one style sandal in this chapter, but once you've learned the basic procedure you can design your own or adapt styles you see around.

Materials

1. Cardboard (thin)
2. Awl
3. Utility knife
4. Stript Ease
5. Edge beveler
6. 1" oblong hole punch
7. Rawhide mallet
8. Dye and saddle soap (optional)
9. Skiving knife
10. Glue (Craftsmen's Cement or Barge Contact Cement)
11. 4/8" sole nails
12. Metal hammer
13. Coarse sandpaper
14. Two ¾" buckles
15. Medium rivets
16. Rivet setter
17. #1 hole punch

You'll need enough 9 to 11 oz. sole leather for two soles (see Suppliers List). The insoles and straps can be made from 5/6 oz. or 7/8 oz. latigo.

Problems

Since the straps on your sandals will loosen up a bit with wear, make them slightly tighter at first. The heel strap can be adjusted by adding more holes.

The sole nails must clinch over the leather in order to hold properly. If they don't, you may need 6/8" rather than 4/8" sole nails. Be sure to hammer them on a metal block or anvil.

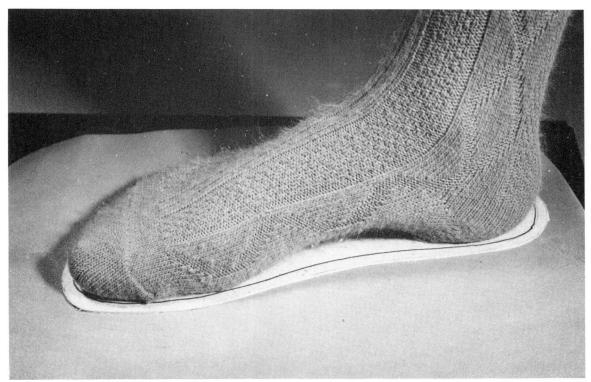

1. Trace your foot on thin cardboard, allowing about ¼" excess all the way around. Cut out the foot pattern. You may round or square off the heel first if you wish.

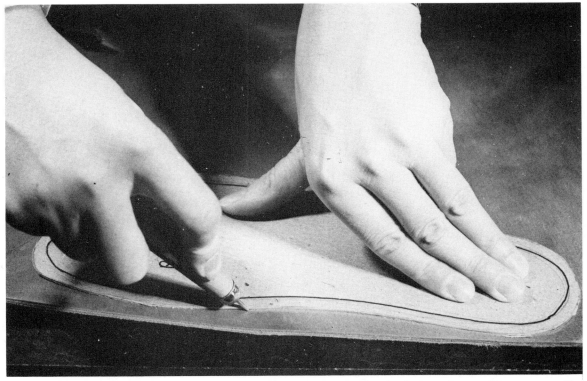

2. With an awl, trace the foot pattern and cut it out with a utility knife from latigo. This piece will serve as the insole.

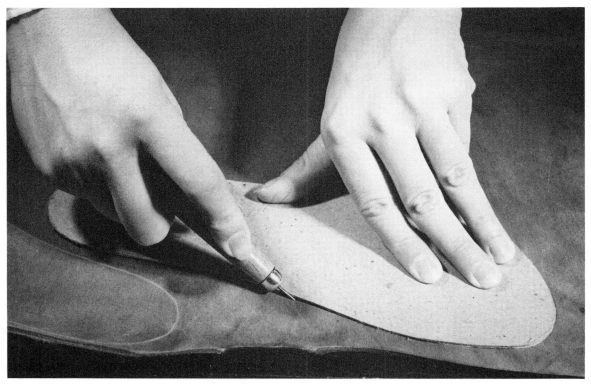

3. Reverse the foot pattern, and trace and cut the other foot (insole) from latigo.

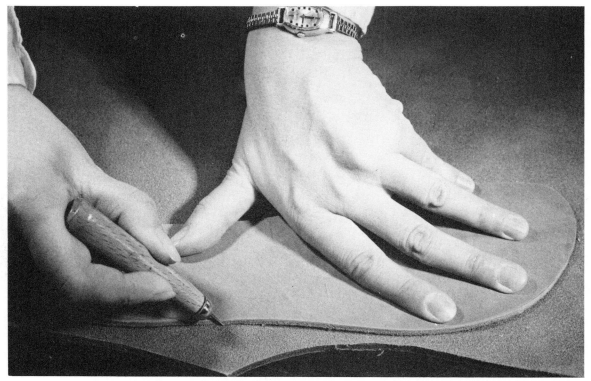

4. Using the latigo insoles as patterns, place them on the *flesh* side of the sole leather. Trace and cut out the holes.

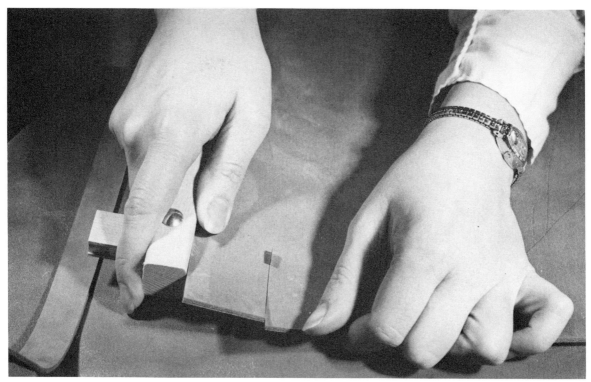

5. Cut two straps of each of the following sizes from latigo with your Stript Ease: 10″ long, 1½″ wide; 12″ long, 2″ wide; 14″ long, ¾″ wide; 4″ long; ¾″ wide.

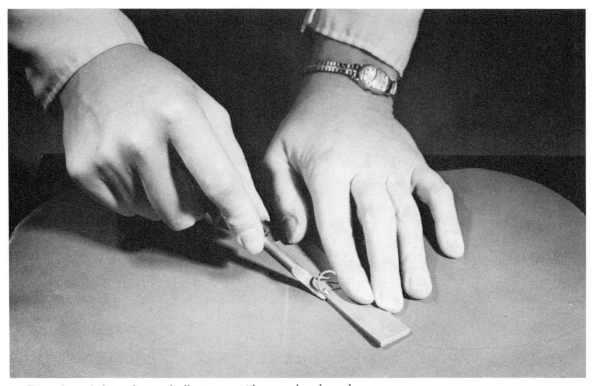

6. Then bevel the edges of all straps with an edge beveler.

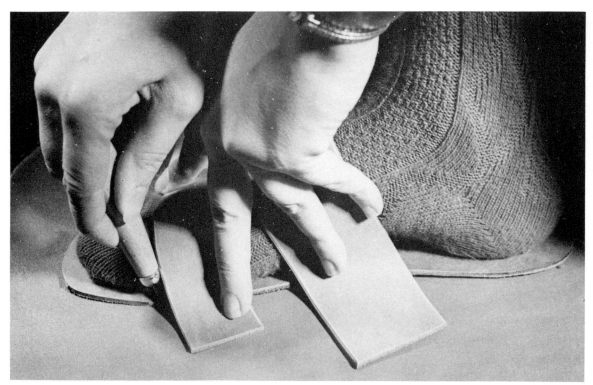

7. Place your foot on the insole and plan the placement of the two front straps (1½" wide and 2" wide). Mark the insole with an awl about 5/16" from the edge to show where the straps will go.

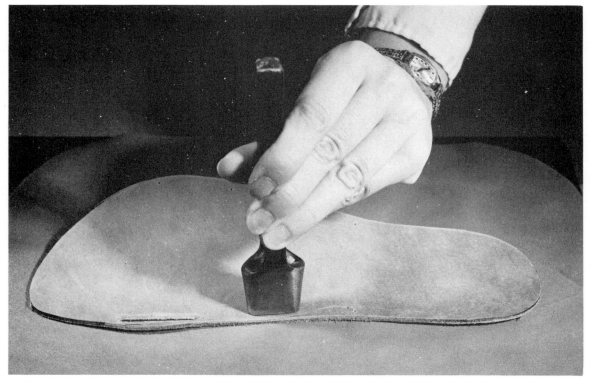

8. Punch the strap slots in the insoles with a 1" oblong punch, about 5/16" from the edge.

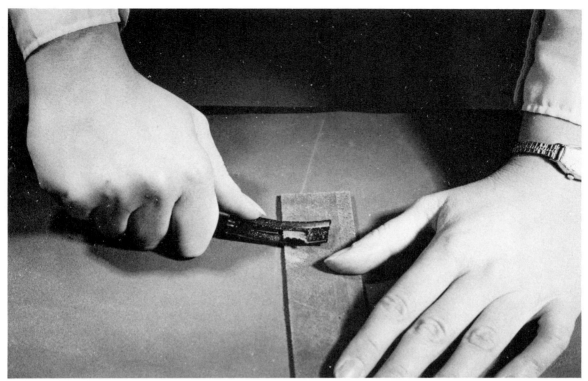

9. Skive the ends of the 1½″ wide and 2″ wide straps where they'll be riveted on the inside. Cut off the ends if they overlap each other underneath. Dye the leather at this point if you wish.

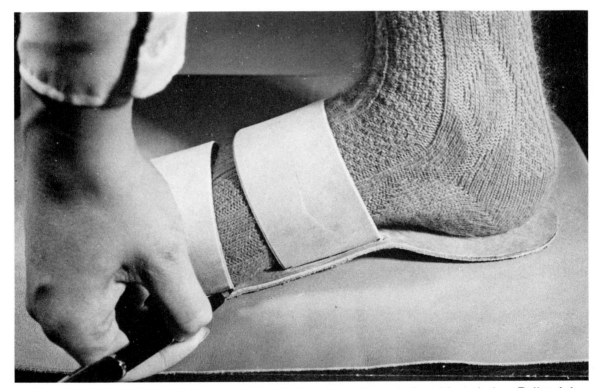

10. Put your foot on the insole again and push the straps through the oblong holes. Pull tightly and mark each strap with a line where it meets the insole.

11. Remove your foot, punch holes, and rivet the straps to the insole with 2 rivets on each side of each strap.

12. Glue the bottom of the insoles to the flesh side of the soles with Craftsmen's Cement or Barge Contact Cement.

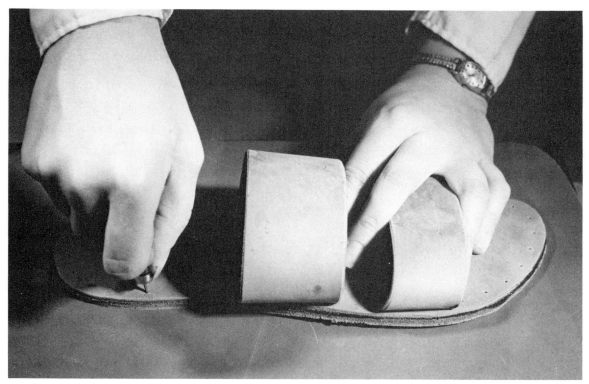

13. When the glue is dry, use an awl to make holes (they don't have to go all the way through) around the outer edge of the sandal and about ¼″ from the edge.

14. Insert sole nails in the holes and nail with a metal hammer against a piece of metal.

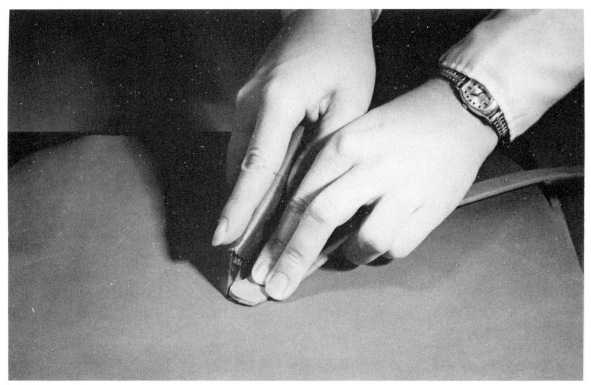

15. Trim the excess leather from the sandal with a utility knife and sand with coarse sandpaper.

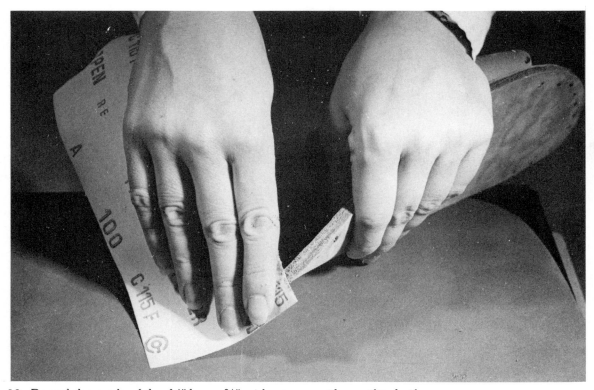

16. Round the ends of the 14" long, ¾" wide straps with a utility knife.

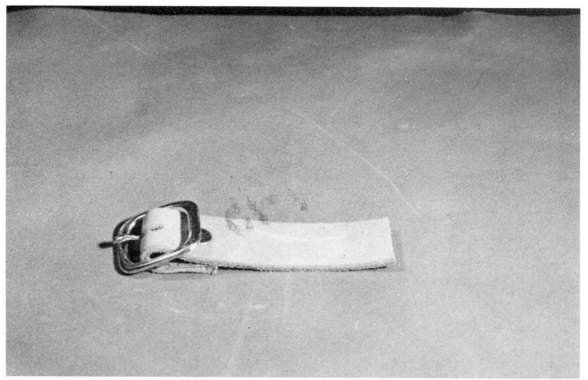

17. Then attach a buckle to the end of each 4″ long, ¾″ wide strap (see Chapter 7 for putting on a buckle).

18. Rivet the straps with buckles to the right side of the right sandal and the left side of the left sandal as shown.

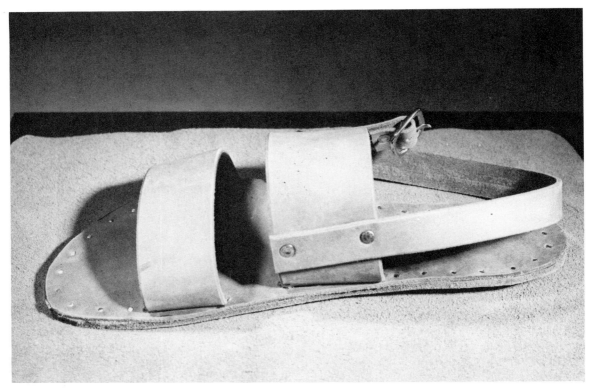

19. Rivet one end of the 14″ long, ¾″ wide strap to the left side of the right sandal and the right side of the left sandal. With your feet in the sandals, mark the comfortable placement of the heel straps and buckles.

20. Punch holes in the heel straps so the sandals will fit comfortably.

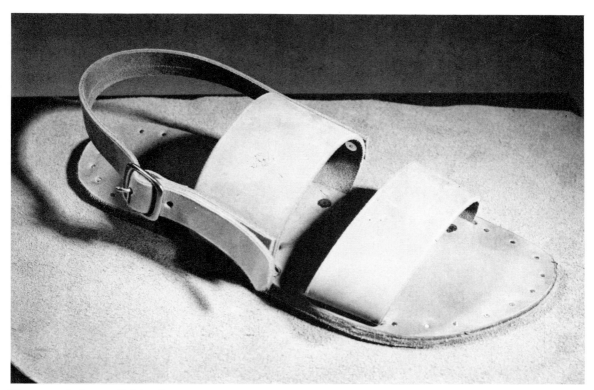

Your sandals are finished.

12 Three-Piece Handbag

Latigo handbags are very popular today, especially among young people, because they are both practical and attractive. You can make them to suit your tastes and needs—large or small, with shoulder or over-the-arm straps, with elaborate decoration or left natural. A leather handbag that lasts and lasts is a popular item in any woman's wardrobe, and lately it is becoming a practical accessory for men, too.

The basic latigo bag shown in this chapter is made with a front, a back, and a gusset. It's an average size and has a shoulder strap.

Materials

1. Awl
2. Utility knife
3. Stript Ease
4. Edge beveler
5. #1, #4 hole punch
6. Rawhide mallet
7. Dye and saddle soap (optional)
8. Rivets
9. Rivet setter
10. Latigo life-eye lacing needle
11. 72" length of ⅛" latigo lace
12. Scissors

For handbags, 7/8 oz. latigo is best. Thinner leather won't hold up as well, and heavier leather is difficult to lace. See the end of this chapter for the handbag pattern.

Problems

If you come out at the end of the bag with a hole left over in one piece of leather, you've most likely either skipped a hole while lacing or punched an extra hole by mistake. Be careful to punch the right number of holes and not to miss any while lacing.

If the bag has puckered excessively after you've finished, you've laced it too tightly. Usually, however, a beginner will not lace tightly enough. In order for the bag to hold together without spaces or air holes, some pucker is necessary. Refer to Chapter 9 for more information on lacing.

1. Trace with an awl and cut each pattern out of latigo. Cut a 38″ long, 1″ wide shoulder strap with a Stript Ease.

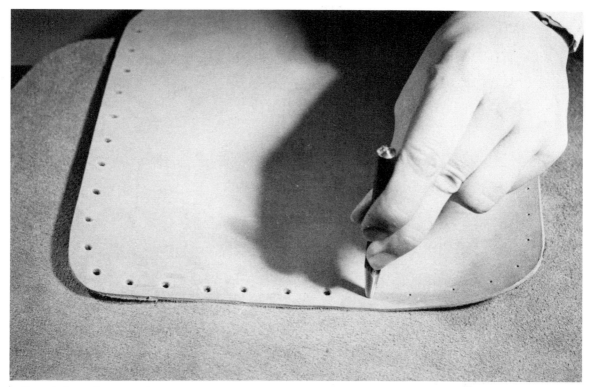

2. Bevel the edges of all pieces with an edge beveler. Mark the lacing holes with an awl and punch them out with a #4 hole punch. You may now dye or decorate the leather if you wish.

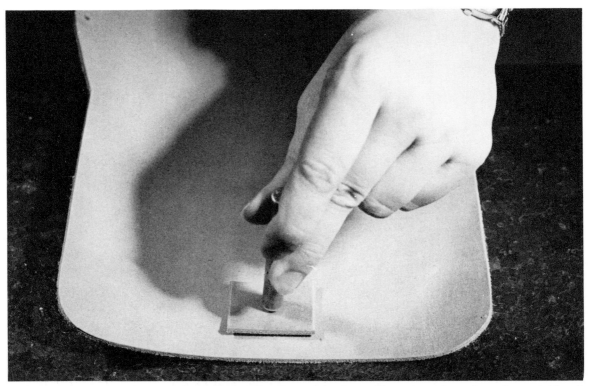

3. Punch a #1 hole through the tab and the flap of the bag. Then rivet the tab to the outside of the flap, as shown.

4. After punching a #1 hole, rivet the shoulder strap to each end of the gusset, with the strap on the *inside*.

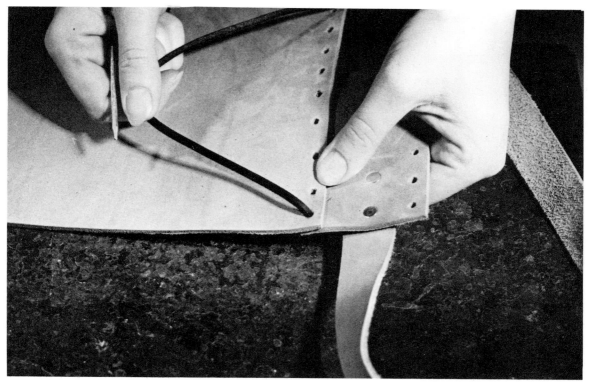

5. With a ⅛″ latigo lace and a latigo life-eye lacing needle, begin lacing from the inside of the gusset through the inside of the bag front (refer to Chapter 9 for detailed lacing instructions).

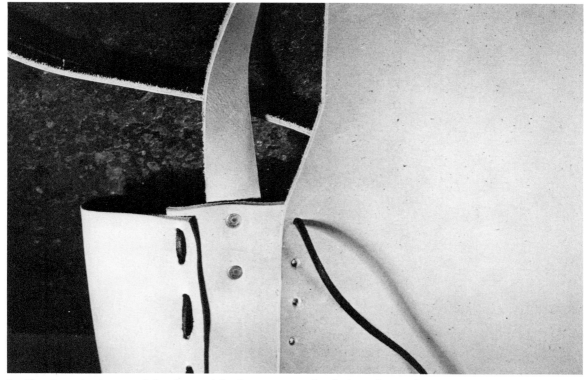

6. Continue lacing until the front of the bag is completely laced; tie a knot, cut, and lace the back of the bag to the gusset.

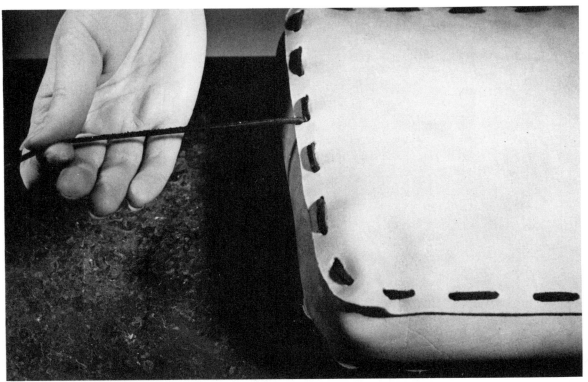

7. After the bag is laced, cut and leave a 6″ piece of lace on the needle. From the front side, push the needle through a bottom hole in the center of the bag and leave the end partway out. Tie a knot on the inside, pulling the lace tightly against the knot.

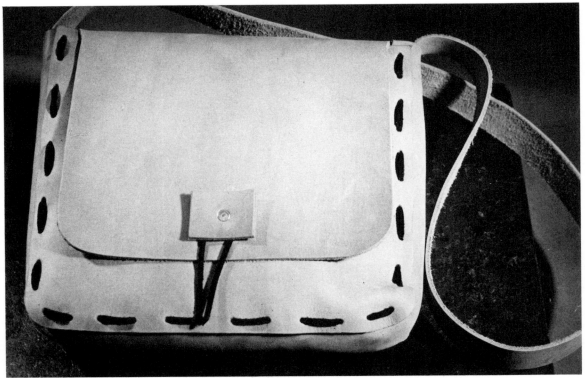

8. Tie this piece of lace around the tab on the front flap for a closing and your bag is completed.

This is a basic three-piece bag with front, back, and gusset. The collage design on the front flap was cobbled with clinch nails. By Lawrence M. Doe of Traverse City, Michigan.

Use this pattern for the front piece of the three-piece handbag. The distance shown by the broken lines should measure 4″ (a total length of 16″). Cut one.

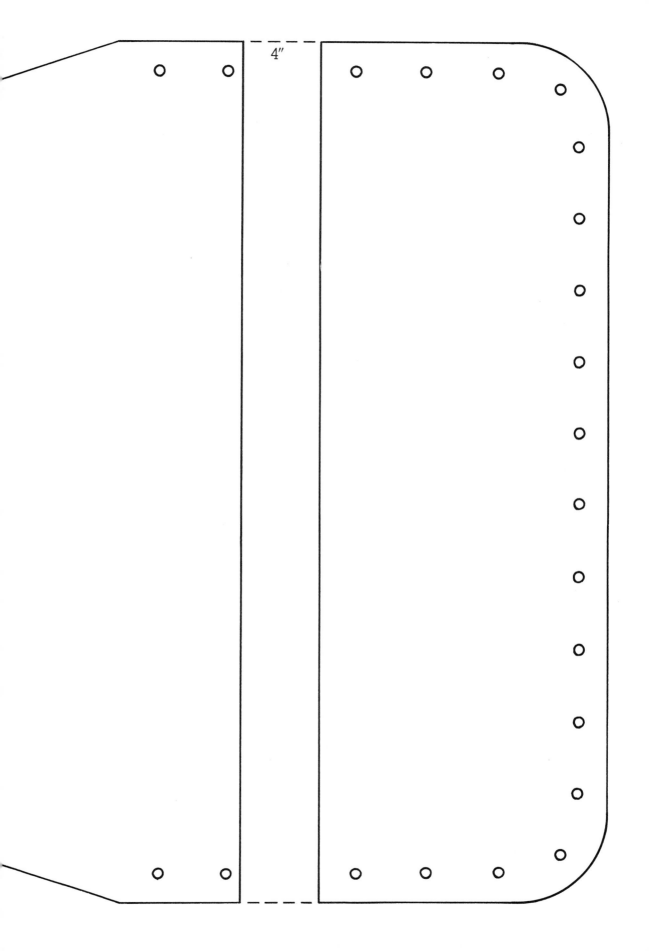

4″

3¼"

(Left) This is a pattern for the front piece of the three-piece handbag. The distance shown by the broken lines should measure 3″ (a total length of 8¼″). Cut one.

10″

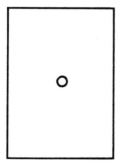

This pattern is the tab of a three-piece handbag. Cut one.

This pattern is the gusset for a three-piece handbag. The distance shown by the dotted line should be 10″ (a total length of 24″). Cut one.

13 Two-Piece Handbag

In Chapter 12 we showed you how to make a handbag with three main parts—the front, the back, and the gusset. In this chapter, we're going to make a handbag that has the front and back combined into one long piece. Instead of a gusset, there are two matching side pieces. Once the pattern is laid out, you will see that this design is more economical in the use of leather. This bag can be made in several styles by varying the shape and width of the sides and main piece.

Materials

1. Awl
2. Utility knife
3. Stript Ease
4. Edge beveler
5. 1" oblong punch
6. #1, #4 hole punches
7. Rawhide mallet
8. Dye and saddle soap (optional)

9. Latigo life-eye lacing needle
10. 60" long piece of ⅛" latigo lace
11. Scissors

Again, we'll be using 7/8 oz. latigo. Use the pattern at the end of this chapter.

Problems

When punching the lacing holes, be sure that the holes opposite the oblong holes are placed halfway between the top and bottom of the oblong holes. Otherwise your front strap won't lace up properly.

If you re having difficulty lacing, refer to the Problems section of Chapter 9.

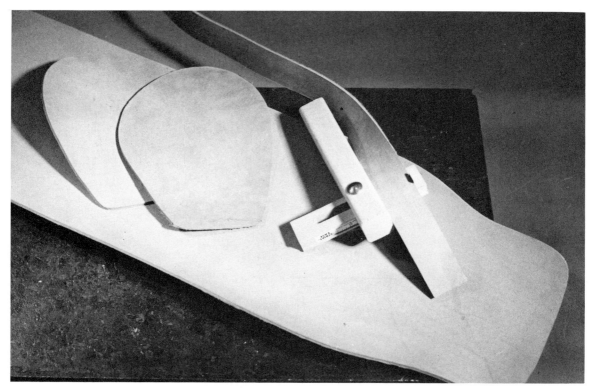

1. Trace each pattern with an awl and cut it out of latigo with a utility knife. Also cut a 36″ long, 1″ wide shoulder strap with a Stript Ease. Bevel the edges of all pieces with an edge beveler.

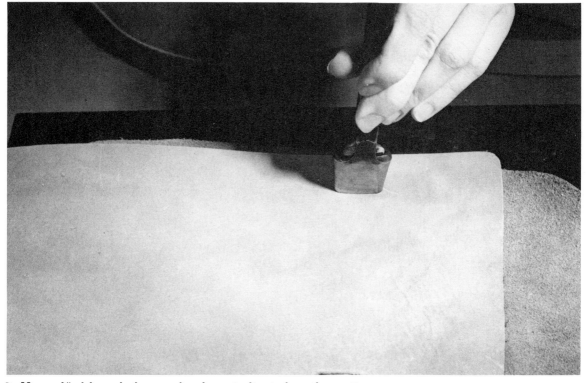

2. Use a 1″ oblong hole punch where indicated on the pattern.

3. Then mark with an awl and punch all lacing holes with a #4 hole punch.

4. Place the shoulder strap end pattern over the strap and mark the four lacing holes on each end. Then punch them out with a #4 hole punch. You may now dye or decorate the leather.

5. Place the front strap through the oblong holes in the bag, as shown, so you won't forget to lace it into the bag.

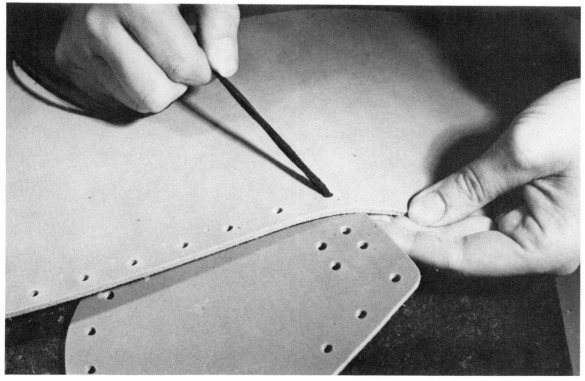

6. Now start lacing from the inside of one side piece through the inside of the top hole in the main piece (refer to Chapter 9 for detailed instructions on lacing).

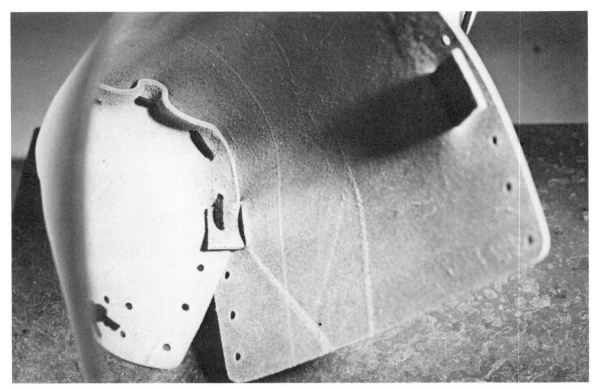

7. Continue lacing tightly around the side of the bag. When you get to the front strap, be sure to lace it in with the other pieces to hold it in place. It should be inside both pieces of leather, as shown.

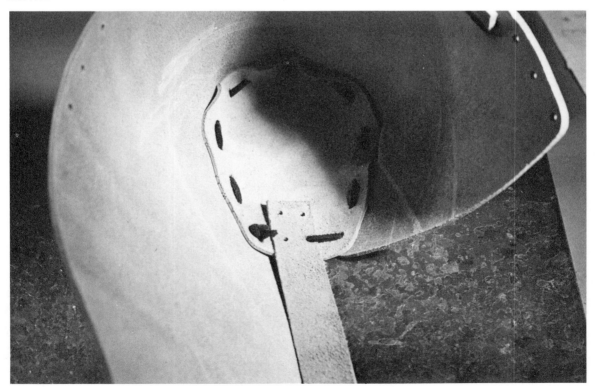

8. When you reach the end, place the shoulder strap against the inside of the bag, matching up the four holes, and push your lacing needle through the closest top hole.

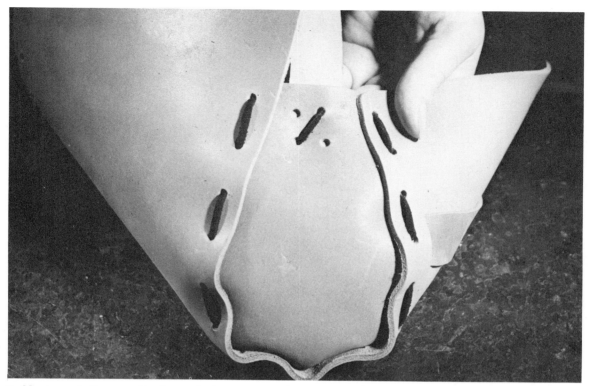

9. Next, push the needle through the opposite bottom outside hole and then through the top inside hole.

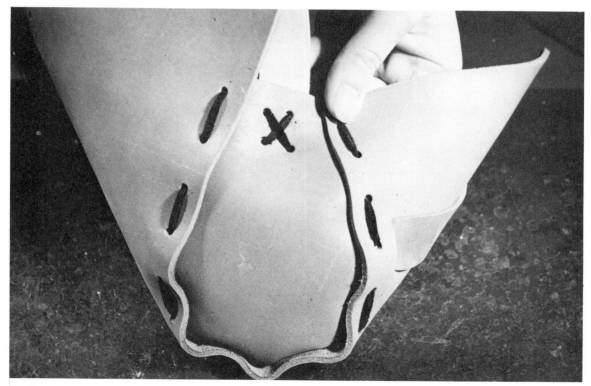

10. Push the needle through the last of the four holes and tie a simple knot tightly against the leather on the inside. Note that you now have an "X" pattern laced on the outside of the bag.

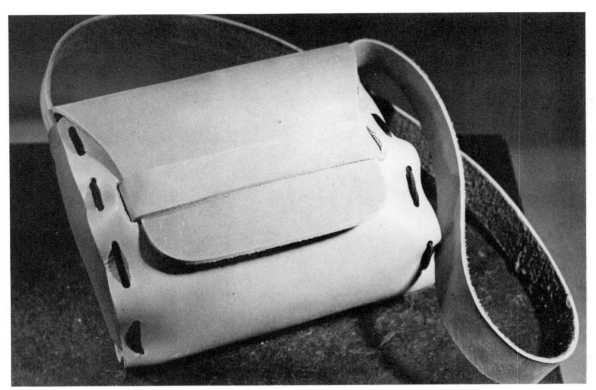

11. Lace the other side of the bag in the same manner. Push the bag flap through the front strap to hold it in place, and you're finished.

33"

Use this pattern to mark where the lacing holes should be punched at each end of the shoulder strap. Cut one 36" long, 1" wide shoulder strap with a Stript Ease, as indicated in Step 1 of this project.

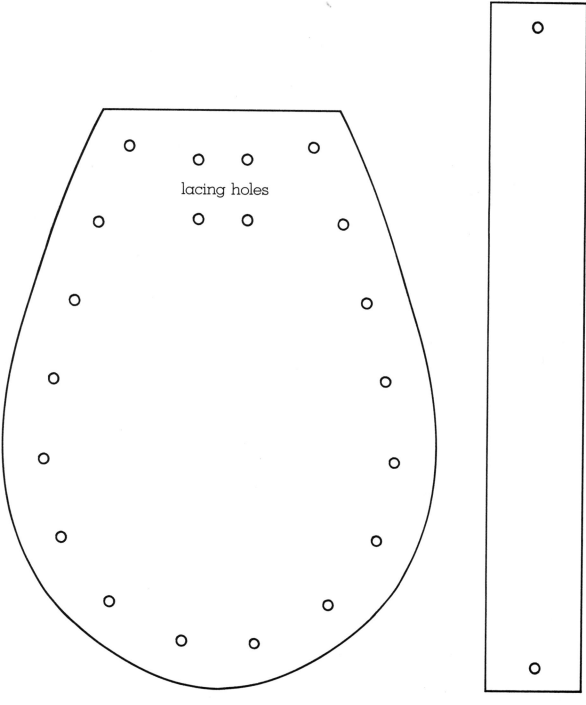

lacing holes

Cut two pieces of this
pattern of the side piece for a
two-piece handbag.

This pattern is the
front strap for a
two-piece hand-
bag variation. Cut
one piece.

9⅛"

This pattern is the main piece of a two-piece hand-bag. The distance shown by the broken lines should measure 9⅛″ (a total length of 21¼″). Cut one.

14 Designing Handbag Patterns

After you've made the two handbags in Chapters 12 and 13, you may want to design your own patterns. This way you'll be able to make a bag the exact size and shape you want. The important thing to remember in designing your own bag is that the greater care and effort you put into the paper pattern, the greater chance there is that the bag will work the first time you make it in leather.

Materials

1. Sketching paper
2. Pattern paper (or brown paper bags)
3. Pencil or pen
4. Scissors
5. String
6. Ruler
7. Paper hole punch
8. Heavy cardboard

Problems

Be very careful when you change the width of the gusset. A wider gusset means you must allow more length in the back of a three-part bag or in the main piece of a two-part bag. If you make the gusset narrower, reverse the procedure.

If the gusset doesn't match the rest of the bag when you finish lacing it together, you may have to add to the pattern or cut some off.

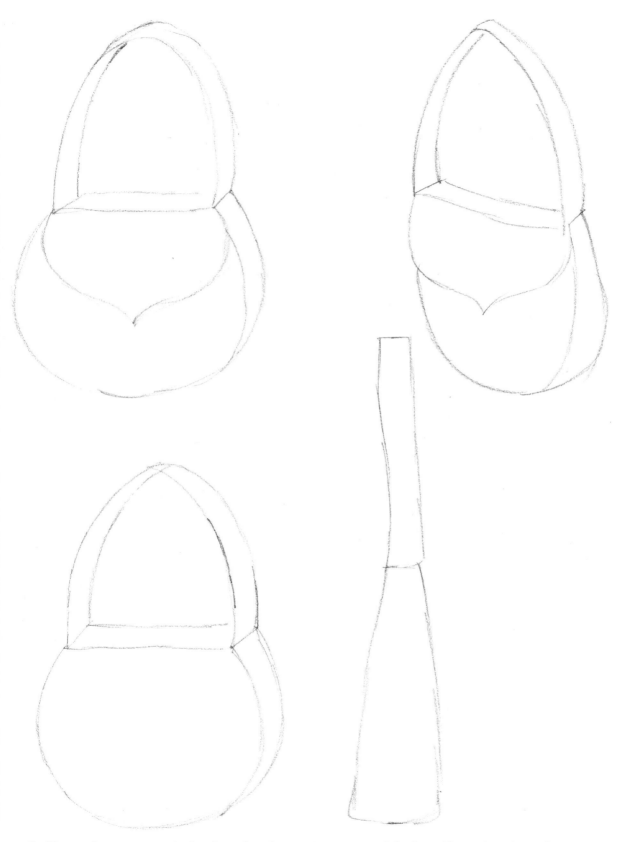

1. First make some rough sketches showing various views of the bag. Try to imagine what your bag will look like from the front, the back, and the sides.

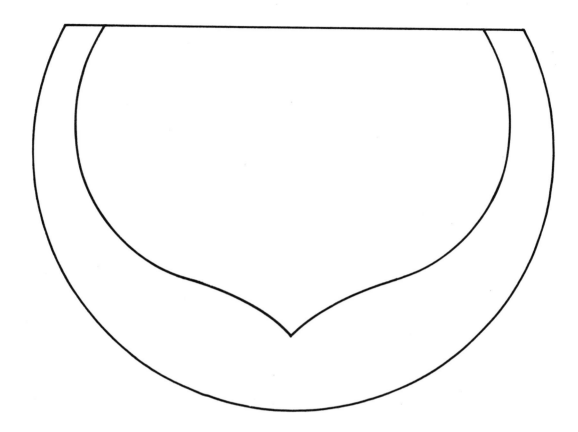

2. On a large piece of paper draw a front view of the handbag the actual size you want it to be.

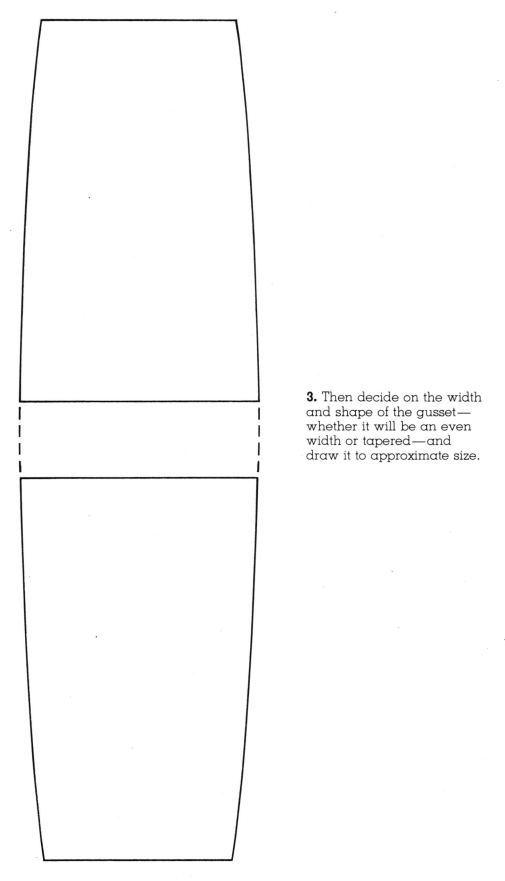

3. Then decide on the width and shape of the gusset— whether it will be an even width or tapered—and draw it to approximate size.

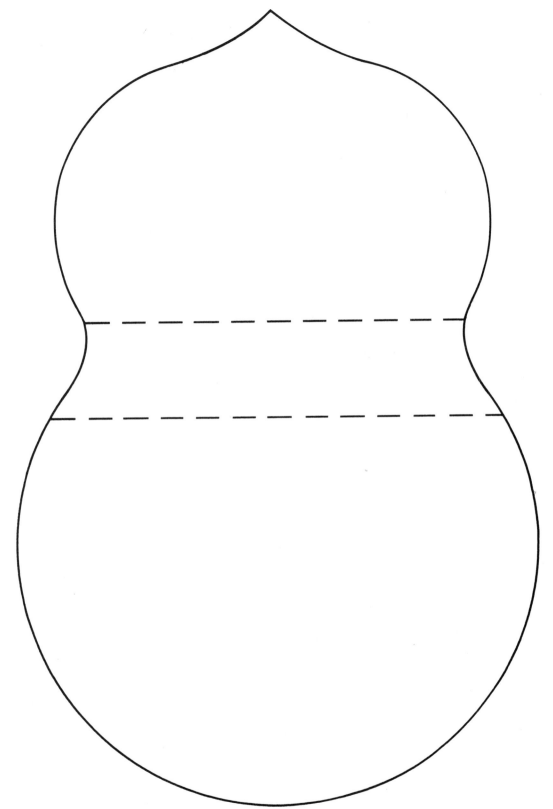

4. Draw the bag back (the actual size), allowing for the bottom of the handbag, gusset room, and the front flap. *Note:* In a three-piece handbag, you'll need a gusset that goes all the way around the bag. The width of this gusset at the top must be added to the length of the back piece. In a two-part bag, add the width of the side pieces to the length of the main piece.

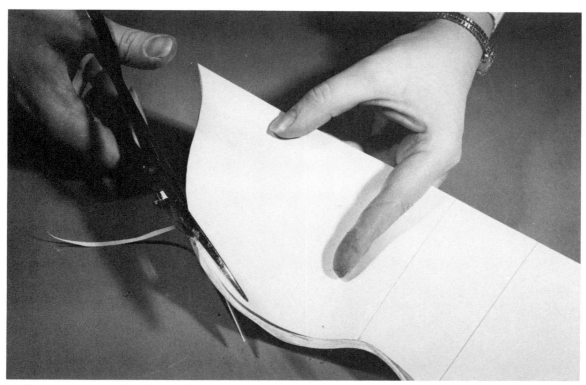

5. Next cut out the bag back, fold it in half lengthwise, and trim it so that both sides are even.

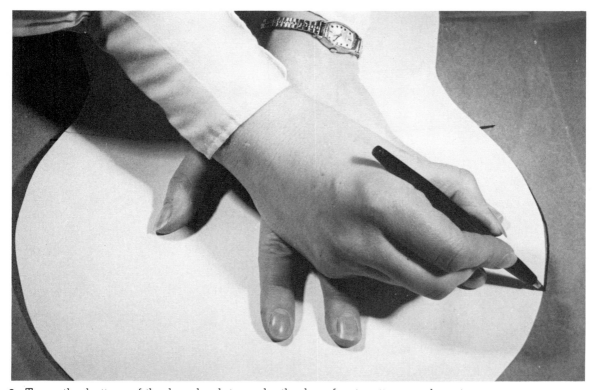

6. Trace the bottom of the bag back to make the bag front pattern and cut it out.

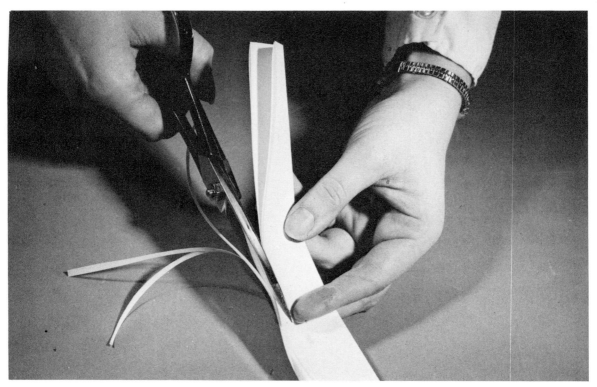

7. Cut out the gusset, fold it in fourths, and trim all uneven edges.

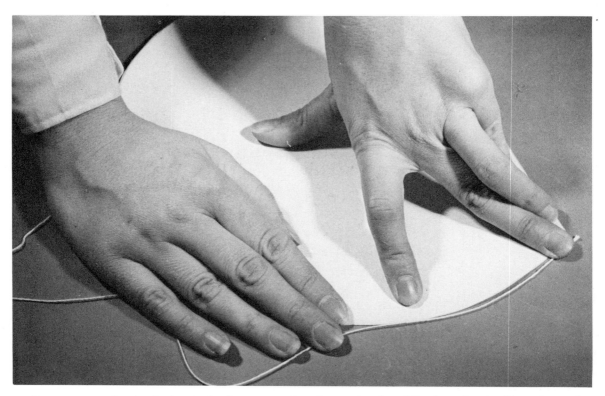

8. Now prepare for the lacing holes by measuring the perimeter of the bag front with a piece of string.

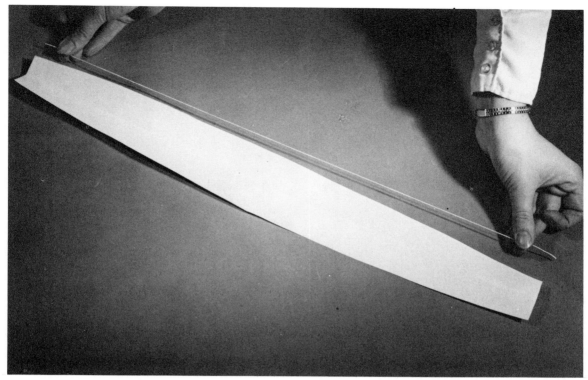

9. Check the gusset length with this string and adjust the gusset if necessary.

10. Draw a line on a piece of paper the length of the string, and starting ¼″ from the end of the line mark off every ¾″. Stop at least 3/16″ from the end.

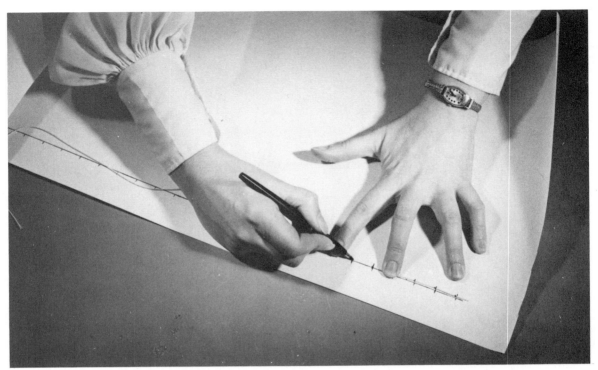

11. There should be an even number of holes in order for the lacing to end up inside the bag. Count your marks. If you have an even number, lay the string along the line and mark it. If you counted an odd number of marks, move the string along the line to eliminate or add a mark and still have between 3/16″ and ⅝″ at each end of the string. If this fails, draw the line again and try marking at ⅝″ or ½″ intervals instead of ¾″. Holes should usually be between ½″ and ¾″ apart.

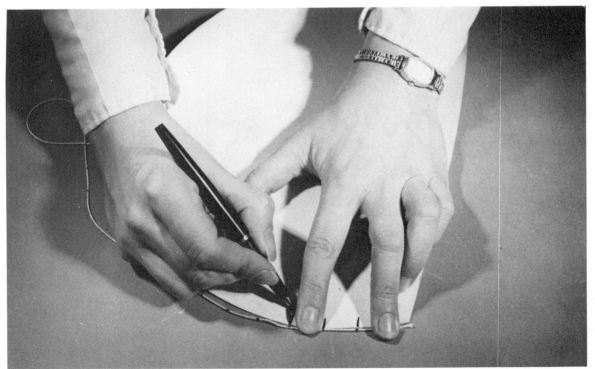

12. When your string is marked, lay it along the perimeter of the bag front and mark off the spaces from the string to the pattern.

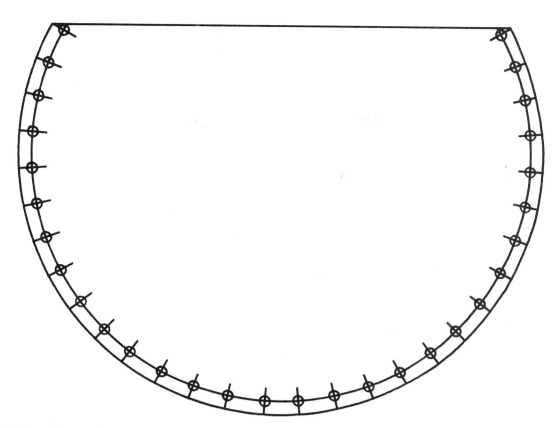

13. Then draw a line around the perimeter of the bag front, ¼″ from the edge. Mark the lacing holes where the lines meet.

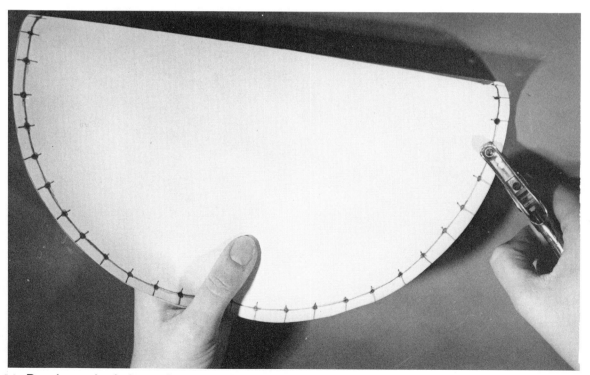

14. Punch out the holes with a paper hole punch.

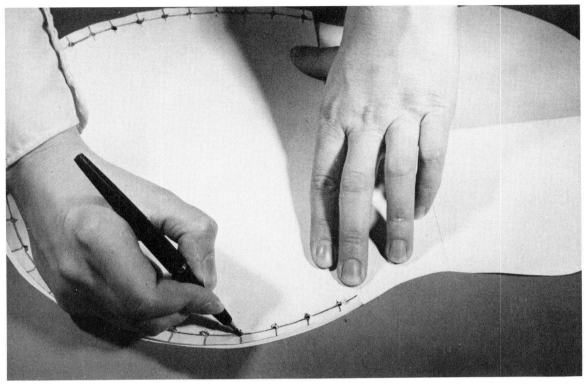

15. Place the bag front over the bag back and punch out the holes.

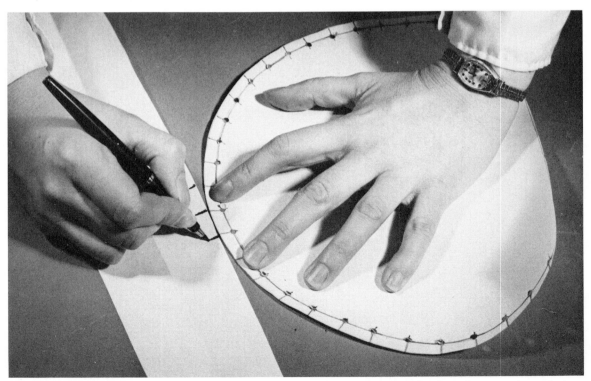

16. Now starting exactly in the middle of the gusset, hold it against the bag front and mark the holes as shown, until you've marked half of the gusset. Then start back at the middle of the gusset and mark the other side.

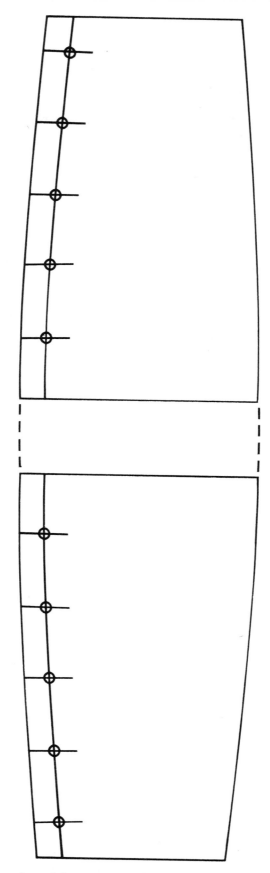

17. Draw a line ¼″ from the edge of the gusset and mark and punch out the holes.

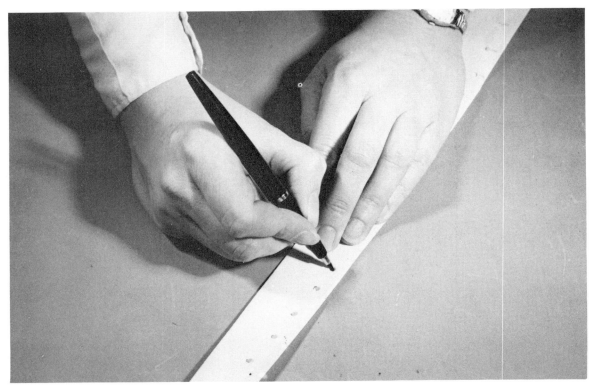

18. Fold the gusset lengthwise, mark the other side, and punch out the holes.

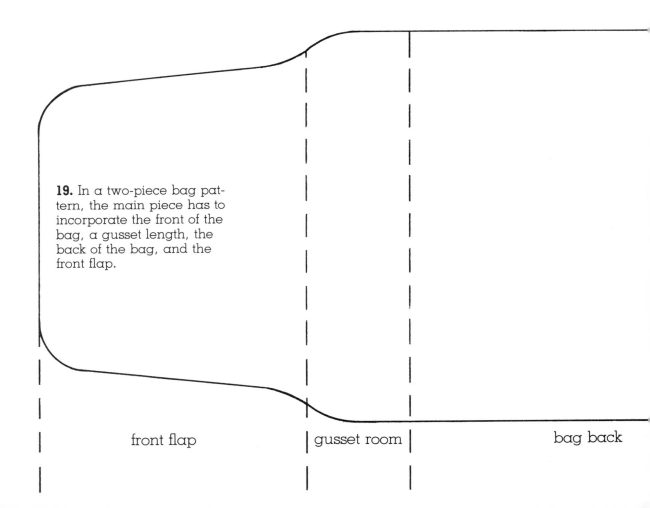

19. In a two-piece bag pattern, the main piece has to incorporate the front of the bag, a gusset length, the back of the bag, and the front flap.

front flap | gusset room | bag back

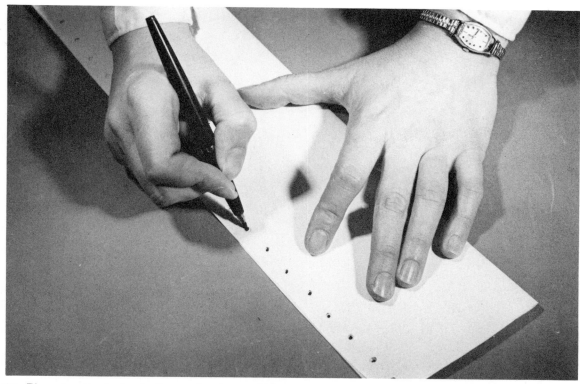

20. Planning the holes in the main piece will be easier on a two-piece bag because the sides of the piece are straight. Do one side and then fold it in half and mark the other side.

| gusset room | bag front |

21. To mark the side pieces, start in the center and work your way around as you did with the three-piece bag. After a thorough rechecking of all parts, your bag pattern is ready for a trial on leather. After making any additional changes from the leather model, copy the bag pattern on heavy cardboard and make a permanent pattern.

This handbag is made from one long piece and two side gusset pieces. By Holy Cow Leather.

This latigo handbag was made from four pieces plus a shoulder strap. By Holy Cow Leather.

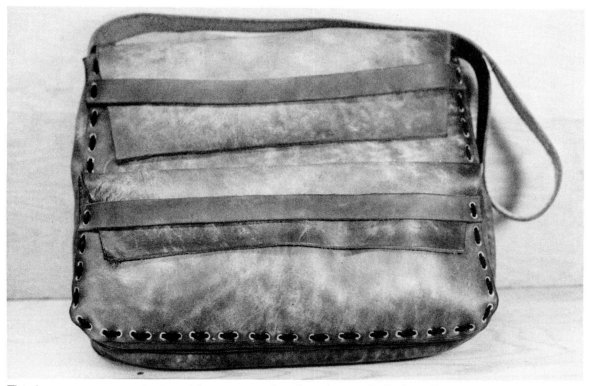

This bag was custom-designed for use as a briefcase/carryall. By Lyn Taetzsch.

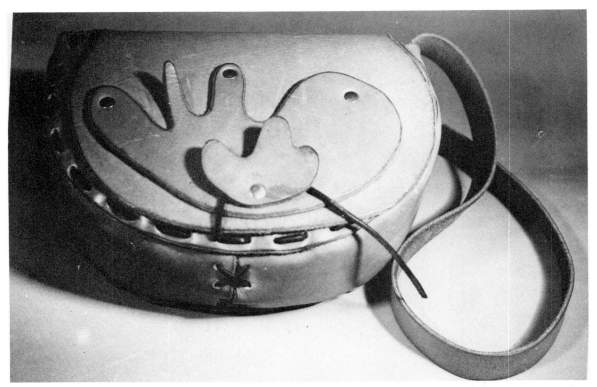

This handbag has a two-piece gusset which is laced on the bottom. The appliqué design was made by riveting pieces of leather to the bag front. By Holy Cow Leather.

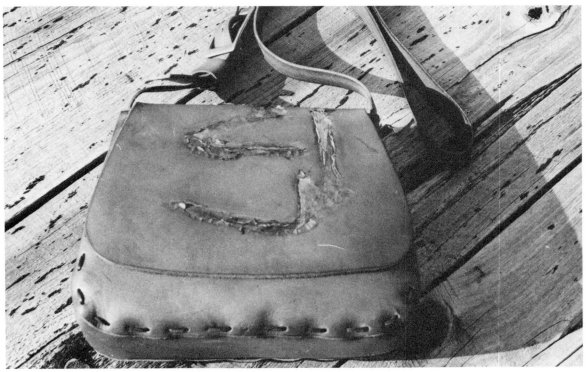

This bag is made from front piece, back piece, and gusset. The design on the front flap is the natural brand which was emphasized with leather dye. By Herb Genfan.

This bag was made from the basic front piece, back piece, and gusset. By Mary Azerbegi of Braum's Leather, Berkeley, California.

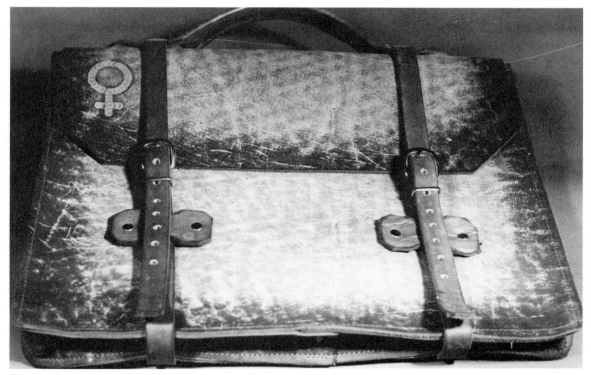

This "schoolbag" type briefcase was made from 6/7 oz. latigo and is machine stitched. If you don't have a heavy-duty sewing machine, hand-stitching will be required on such heavy leather. By Holy Cow Leather.

Suppliers List

Here's a list of suppliers from whom you can obtain all the materials mentioned in the preceding chapters. After you've looked up the suppliers you wish to contact, you'll find their full addresses on the opposite page. Of course, you should also check out your local art supply store, hardware store, shoe repair shop, and scrap steel yard.

Buckles
Berman Leather Co.
California Crafts Supply
Just Brass, Inc.
Skil-Crafts
Tandy Leather Co.
Trinity Buckle Co.

Daubers
Berman Leather Co.
Fiebing Chemical Co.
MacLeather Co.
Skil-Crafts
Tandy Leather Co.

Dye
Berman Leather Co.
California Crafts Supply
Fiebing Chemical Co.
MacLeather Co.
Skil-Crafts
Tandy Leather Co.

Harness Dressing
Fiebing Chemical Co.

Jumprings and Earwires
Bergen Arts & Crafts
Skil-Crafts

Keyrings
California Crafts Supply
MacLeather Co.
Tandy Leather Co.

Laces
Berman Leather Co.
Tandy Leather Co.

Leather: English Kip and Sole
Berman Leather Co.
California Crafts Supply
MacLeather Co.
M. Siegel Co., Inc.

Leather: Latigo
Berman Leather Co.
California Crafts Supply
Hermann Oak Leather Co.
MacLeather Co.
M. Siegel Co., Inc.
Tandy Leather Co.

Leather Cement
Berman Leather Co.
California Crafts Supply
MacLeather Co.
Tandy Leather Co.

Leather Tools
Berman Leather Co.
California Crafts Supply
MacLeather Co.
M. Siegel Co., Inc.
Skil-Crafts
Tandy Leather Co.

Legs and Bolt for Stool
Tandy Leather Co.

Rivets and Snaps
Berman Leather Co.
California Crafts Supply
MacLeather Co.
M. Siegel Co., Inc.
Skil-Crafts
Tandy Leather Co.

Saddle Soap
Berman Leather Co.
California Crafts Supply
Fiebing Chemical Co.
MacLeather Co.
M. Siegel Co., Inc.
Tandy Leather Co.

Sheepskin Scraps
Berman Leather Co.
Tandy Leather Co.

Sole Nails
Berman Leather Co.
California Crafts Supply
MacLeather Co.
M. Siegel Co., Inc.
Tandy Leather Co.

Addresses

Bergen Arts & Crafts
P.O. Box 381
Marblehead, Massachusetts 01945
Catalog $1.00

Berman Leather Co.
145–147 South Street
Boston, Massachusetts 02111
Free Catalog

California Crafts Supply
1096 North Main Street
Orange, California 92667
Free Catalog

Fiebing Chemical Co.
516 South Second Street
Milwaukee, Wisconsin 53204
Wholesale Only

Hermann Oak Leather
4050 North First Street
St. Louis, Missouri 63147

Just Brass Inc., a division of
Richter Bros.
1612 Decatur Street
Ridgewood, New Jersey 11227
Catalog $1.00

MacLeather Co.
424 Broome Street
New York, New York 10013
Free Catalog

M. Siegel Co., Inc.
186 South Street
Boston, Massachusetts 02111
Free Catalog

Skil-Crafts, a division of
Brown Leather Co., Inc.
305 Virginia Avenue
P.O. Box 105
Joplin, Missouri 64801
Catalog $1.50

Tandy Leather Co.
300 Fifth Avenue
New York, New York 10001
Free Catalog

Trinity Buckle Co.
P.O. Box 5169
Santa Monica, California 90405
Free Catalog

METRIC CONVERSION CHART

	Multiply by	From To	To From	Multiply by
LENGTH	0.03937	Inches	Millimeters	25.4
	0.3937	Inches	Centimeters	2.54
	39.37	Inches	Meters	0.0254
	3.2808	Feet	Meters	0.3048
	1.0936	Yards	Meters	0.9144
AREA	0.155	Square Inches	Square Centimeters	6.4516
	10.764	Square Feet	Square Meters	0.0929
	1.196	Square Yards	Square Meters	0.83613
WEIGHT	0.035274	Ounces, Avoirdupois	Grams	28.350

Here are some of the most commonly used fractions:

Inches	Millimeters
1/64	0.40
1/16	1.59
1/8	3.18
1/4	6.35
3/8	9.53
1/2	12.70
5/8	15.88
3/4	19.05
7/8	22.23
1	25.40

Index

Lyn Taetzsch, a skilled leathercrafter, was owner and president of Holy Cow Leather, in Newfield, New York, for three years. She is also a painter, whose work has been shown in galleries throughout the country, including a one-woman show at the Paula Insel Gallery in 1965. Ms. Taetzsch received her B.A. in English from Rutgers University and studied fine art at Cooper Union and the University of Southern California.

Herb Genfan is a professor of Business Administration at Ithaca College, Ithaca, New York. He was the consulting vice-president of Holy Cow Leather for three years. He was the Director of Manpower Development for Blue Cross/Blue Shield in Newark, New Jersey; Organization Analyst for Prudential Insurance; and Organization Planner for Macy's in New York City. He received his B.A. in Psychology from Columbia and his M.A. in Counseling from New York University, where he also did doctoral work in Administration.

Edited by Joan E. Fisher
Designed by Jim Craig
Set in 12 point Stymie Light by Gerard Associates/Graphic Arts, Inc.
Printed and bound by Interstate Book Manufacturers, Inc.